STUDENT UNIT GUIDE

# AS Geography UNIT 1

Edexcel

Specification B

## Unit 1: Changing Landforms and Their Management

Bob Hordern & Sue Warn

Philip Allan Updates
Market Place
Deddington
Oxfordshire
OX15 0SE

tel: 01869 338652
fax: 01869 337590
e-mail: sales@philipallan.co.uk
www.philipallan.co.uk



ISBN 0 86003 468 2



This Guide has been written specifically to support students preparing for the Edexcel Specification B AS Geography Unit 1 examination. The content has been neither approved nor endorsed by Edexcel and remains the sole responsibility of the authors.

Typeset by Magnet Harlequin, Oxford
Printed by Information Press, Eynsham, Oxford

P00039

# Contents

# Introduction

## About this guide

The purpose of this guide is to help you understand what is required to do well in **Unit 1: Changing Landforms and Their Management**. (The full content details are in the Edexcel specification, on pages 18–21. Your teacher or lecturer will have a copy of this.)

The guide is divided into three sections.

This **Introduction** explains the structure of the guide and the importance of finding linkages between river and coastal environments, and between the AS and A2 parts of the course. It also provides some general advice on how to approach the unit test.

The **Content Guidance** section sets out the *bare essentials* of the Unit 1 specification, which comprises two sub-units — river environments and coastal environments. A series of diagrams is used to help your understanding; some are simple to draw and could be used in the exam.

This section helps you develop essential skills, such as responding to data and using case studies. The review exercises help you to test your understanding of each section. There are also some tips on how to answer these questions.

The **Question and Answer** section includes three sample exam questions in the style of real papers. Sample answers at C and A grade are provided, as well as examiner's comments on how to tackle each question and on where marks are gained or lost in the sample answers.

## Linkages

You should try to make linkages between the river environment and coastal environment sub-units.

- There are **physical** linkages, in that rivers feed into the coastal ocean systems.
- There are **parallel** linkages — for example, both systems are in a state of **dynamic equilibrium** and this can be upset by dramatic events such as storms and floods.
- There are **process** linkages, in that similar processes of erosion, transport and deposition operate to produce short-term and long-term changes in landforms. These changes are possibly more concentrated and dramatic on the coast, as it is a much **narrower** zone.
- Both the coast and rivers experience the impact of long-term changes over geological time — changes of sea level have a significant impact on the landforms of both coastal and inland areas.

AS Unit 1 also forms a foundation for A2 study. As part of A2, the final **Unit 6** is a *synoptic exercise*. You are given a series of resources about a geographical issue within an area and, under exam conditions, you have to use these resources to carry out a structured inquiry, considering a number of planning choices for the area. Clearly, if the resources are for a river basin, or an island with a coastal zone, you need to revisit Unit 1.

As the synoptic exercise is testing your ability to draw together what you have learnt (knowledge, understanding and skills) from all the AS/A2 units, you need to be making these linkages throughout the course.

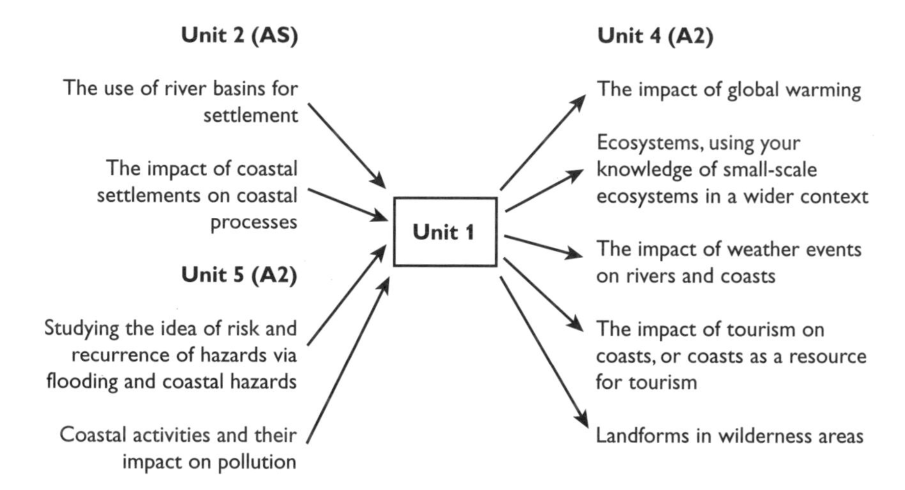


# General advice on AS unit tests

## Choice of questions

You must answer any three of the five questions set. To help you find them, topics will appear in the same order as in the specification: (1) and (2) are questions on rivers or hydrology, (3) is a mixed question, and (4) and (5) are questions on coastal environments.

Always look at the whole question before making your final choice. The mini-essay at the end of each question is worth 10 marks (33%), so don't make your choice based only on the resources, or you may get a shock.

## Timing

The examination lasts for 1 hour 30 minutes. Each question is worth 30 marks, so you have to earn 'a mark every minute'. If you can't answer a subsection, just miss it out and go straight on. You can go back later.

You must answer three complete questions. For the short answers, bullet points are quite acceptable, but the mini-essay expects extended writing. However, even here, notes are better than nothing. Try to write detailed, factual prose, and not to waste words. Make every word count.

## Quality of written communication

For both the AS examination papers, up to 4 marks are added on according to:

- the quality of your spelling (if you are dyslexic, seek special consideration)
- your punctuation and grammar
- the structure and ordering of your response into a logical answer
- appropriate use of geographical terminology

### Concepts and geographical terms

It is a good idea to compile a list of key ideas and terms as you meet them. You might refer to other books (your teacher will have a vocabulary list in their *Teacher's Guide*) and there are many such terms throughout this book (in bold type).

## Managing questions

Go through each subsection lightly underlining or ringing key words. In particular, look for:

- command words (e.g. 'describe', 'explain', 'examine')
- locations (e.g. is LEDC or MEDC specified?)
- the number of reasons or answers required (e.g. *two* ways, *one* reason, etc.)
- 'with the aid of a diagram' — this means you must draw one if you want full marks

Judge the length of your answer. The lines provided on your paper are a guide for writing of an average size. You can write more if you wish, with no penalties, and don't be afraid to draw a diagram — it doesn't matter that there are lines on the paper.

# Content Guidance

Unit 1 comprises two sub-units:

- **river environments**
- **coastal environments**

Note that environments include selected small-scale ecosystems.

In this Content Guidance section, each of these sub-units has the following structure:

- an introduction
- key physical processes and how they lead to short- and long-term change
- people–environment interactions — this is an applied geography section where you explore planning and management issues
- looking ahead to the future — in particular, reviewing the feasibility and economic viability of more sustainable strategies

# Introducing fluvial systems

## Understanding the hydrological cycle

The **global hydrological cycle** is the continual circulation of water between the Earth's surface and the atmosphere. It is an example of a **closed** system, whereby water circulates through a number of **stores**. The **output** from one store therefore becomes the **input** to the next, as a result of a series of **flows** brought about by processes such as condensation, precipitation and evaporation. Water moves through the cycle in all three forms — as a liquid, a solid (ice) and a gas (water vapour).

In terms of total volume, the Earth has plenty of water, especially when compared with other planets. However, only a small percentage of this water — perhaps 8% — is readily available for use. The rest is stored in saline oceans, or in polar ice caps, or deep below the land's surface.

Nor is this water evenly distributed over the Earth's surface. The hot, dry tropics and interiors of some continents are very deficient in water, in contrast to equatorial and temperate mid-latitudes. Where there is an enormous and rapidly increasing demand, as in California, this can have a major impact on rivers, such as the Colorado, which barely reaches the sea because of the huge amounts of water extracted from it.

As the global population continues to grow, and demand for agricultural and industrial use of water escalates, there is a fear that scarcity of water could even lead to wars between countries in areas of deficit, such as the Middle East.

Table 1 shows the imbalance of supply between selected countries.

***Table 1 Renewable fresh water resources of selected countries, 1995***

| Country | Total (km$^3$ year$^{-1}$) | Per person (thousand m$^3$ year$^{-1}$) |
|---|---|---|
| Brazil | 5190 | 32.12 |
| Canada | 2901 | 109.51 |
| Egypt | 2 | 0.04 |
| Norway | 405 | 97.10 |
| Pakistan | 298 | 2.32 |
| UK | 120 | 2.14 |
| Source: UNEP (1996) | | |

### Review 1

Outline the physical and human factors that cause the amount of water per person to vary in the countries shown in Table 1.

***Tip*** Think of supply and demand.

A further dimension to the global hydrological situation is brought about by **global warming**, which could play havoc with established patterns of water supply. Today's 'wet' areas with few water supply problems, such as southeast England, could become much hotter and drier in summer, with drought a regularly occurring hazard.

Water planners will therefore need to consider a number of ways in which they can increase water supplies in areas of shortage — for example, by:

- increasing the number of artificial storage reservoirs (e.g. in the Deccan in India)
- exploiting even more and deeper groundwater supplies (e.g. in the Beijing area in China)
- using more of the ocean's saline water by developing desalinisation plants (e.g. in Saudi Arabia)
- diverting waters from snow/glacier stores to areas of deficit (e.g. in the western USA)

However, these engineering works take place at huge economic and environmental cost, so water planners are increasingly looking at more **sustainable** use of water — for example, by **water conservation** and **recycling** of water for agricultural and industrial use, where quality is less of an issue.

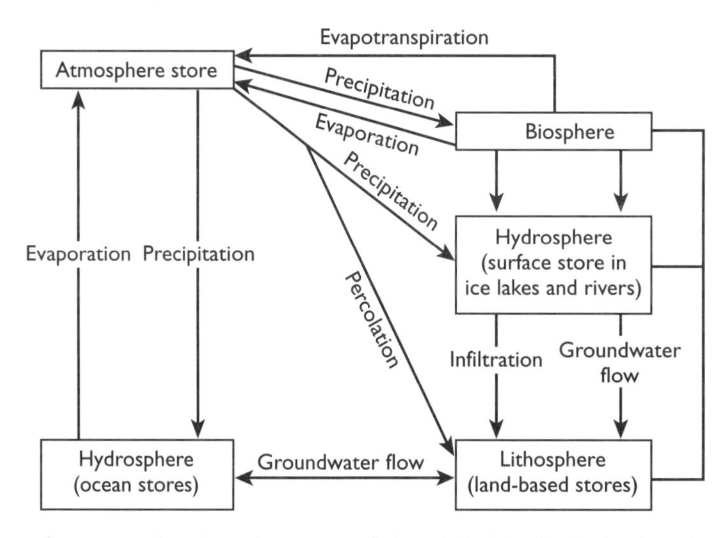


***Figure 1 The closed system of the global hydrological cycle***

# The drainage basin system

The **drainage basin system** is an example of an open system within the closed system of the global hydrological cycle. You can think of the drainage basin system as a black box with a number of processes operating in it. These processes can impact on river flow and the soil, and also on how sediment is transferred within the drainage basin.

***Table 2 The likely impact of factors on processes in the drainage basin system***

| **Factor** | **Runoff** | **Sediment** |
|---|---|---|
| Rock type | Permeable rocks (e.g. limestone) act as groundwater reservoirs (i.e. aquifers), leading to a steady runoff. Impermeable rocks (e.g. clay) respond rapidly to precipitation — flashy hydrographs with peaks and lows | Unconsolidated sands and gravels supply much greater amounts of sediment than hard igneous rocks or limestone, which is corroded in solution and supplies little sediment |
| Climate | The amount of precipitation is a major factor in influencing runoff. Temperature has a major impact on amounts of evaporation. High temperatures will reduce runoff because of evaporation | Precipitation can lead to rapid runoff and intense sedimentation. Duration, intensity and seasonality are all important. The nature of the climate has a major impact on the type of weathering and mass movement (i.e. the sediment supply) |
| Slope, topography and altitude | Steep slopes will lead to rapid runoff. Altitude will have an indirect effect on precipitation | Steep slopes will lead to active mass movement with rapid runoff, causing rill erosion and large inputs of sediment |
| Soils | The nature of the soil will influence infiltration rates (e.g. where soil is baked hard by the sun, infiltration rates are low and runoff will show a rapid response) | Particle size, cohesiveness and infiltration capacity will all influence amount of sediment supply |
| Vegetation | Vegetation intercepts the rainfall and cuts down the amount of water reaching the river as runoff | The amount and nature of vegetation cover will determine the degree of protection from soil erosion. Vegetation binds the soil together |
| Land use | The type of land use has a major impact on infiltration rates. Where grasslands are badly eroded, this can lead to a lot of direct overland flow. Levels of urbanisation control the speed at which rainfall reaches the system. Impermeable surfaces lead to high amounts of runoff | Some land uses, such as downslope ploughing, promote rill erosion. Except where construction is occurring, highly urbanised areas yield lower amounts of sediment |
| Size and shape of basin | Generally, larger catchments generate more runoff if other factors are equal. The shape of basin can have a major impact on the pattern of response at a particular point along a river | Clearly, larger basins yield more sediment, but often other factors are more important, such as rock type |
| Drainage density | High density of runoff can lead to high levels of runoff | High drainage density encourages soil erosion, but irrigation channels can encourage siltation etc. and cause less sediment to reach the river |
| Human activities | Dams can control variability in runoff | Dams can act as an 'on-river silt trap', with sediments piling up in the lake behind the dam |

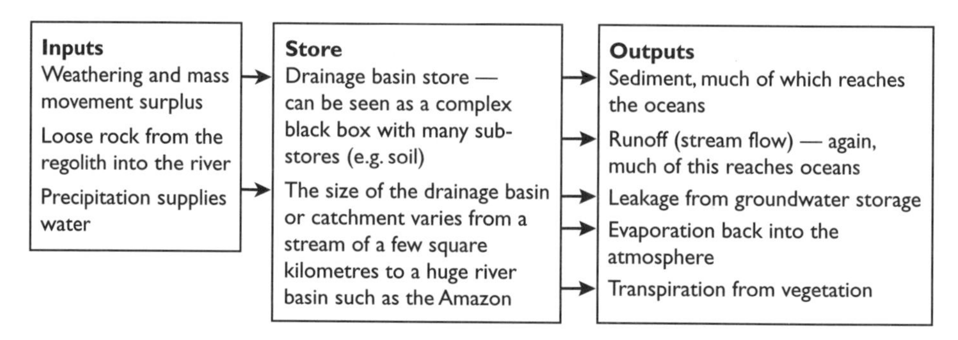


***Figure 2 An outline of the drainage basin system***

The balance between the water inputs and water outputs is known as a **water budget** (see Figure 3).

Whatever the size of the drainage basin, a number of **factors** affect the drainage basin system to a lesser or greater degree (see Table 2, page 11).

# Water budget models

The water balance available in the soil is of vital importance to users such as farmers, who can use it to identify the best times for irrigation, etc. You can see how this varies over a year by plotting temperature, precipitation and potential evapotranspiration on a water budget graph.

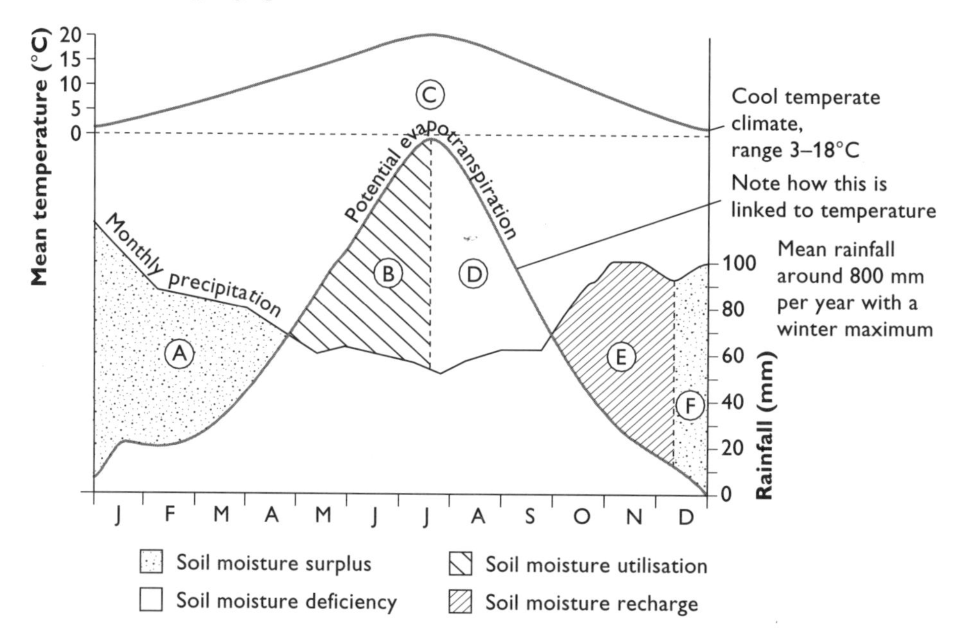


***Figure 3 A water budget graph for a cool temperate area in south Wales***

The sequence below shows you how to interpret the water budget graph.

**A** Precipitation is greater than potential evapotranspiration. The soil water store is full, so there is a surplus of soil moisture for plant use, runoff into streams and recharging groundwater supplies.

**B** Potential evapotranspiration is greater than precipitation, so plants must rely on the water stored. It is gradually used up.

**C** Losses by evapotranspiration and usage by plants mean that the soil moisture store is now used up; only irrigation can replenish supplies of water for crops. River levels fall disastrously. If there are storms, direct overland flow may occur over the baked ground, or if infiltration operates effectively, all the water will be absorbed in the ground.

**D** This is the period of soil moisture deficiency. Plants must have adaptations to survive if it is a long period, and crops must be irrigated.

**E** Precipitation is now greater than potential evaporation, so the soil water store will start to fill up again. If the rains do not come, this can be a very problematic period.

**F** The soil water store is now full (i.e. **field capacity** is reached); any surplus water will percolate down into the groundwater store.

Water budget models can be drawn for a variety of climatic types. The relationship between precipitation, temperature and potential evapotranspiration can have a major impact on the sequence **A–F**.

# River regimes

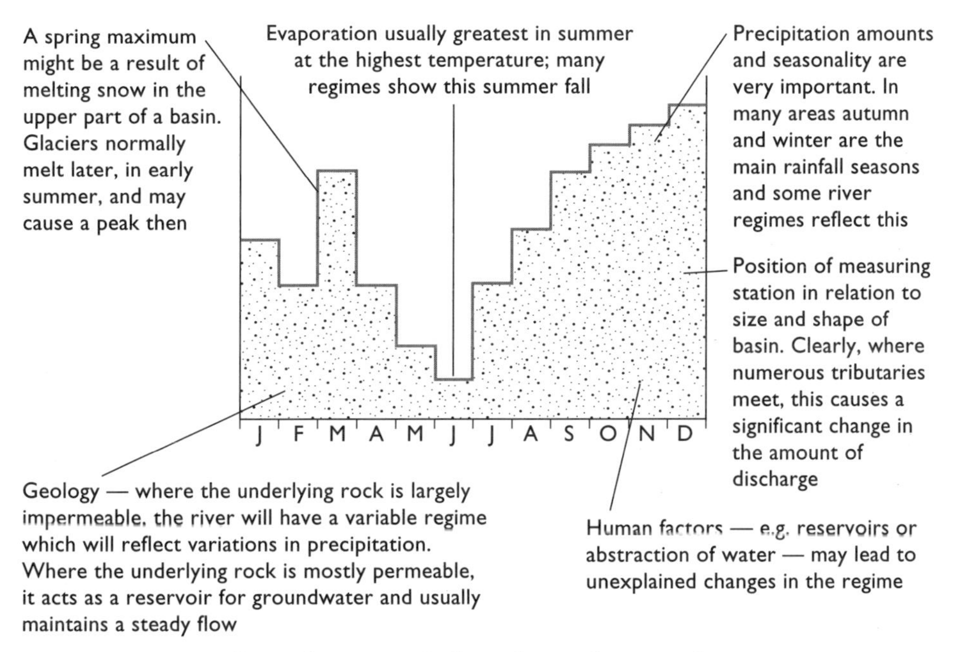


***Figure 4 Factors influencing a river's regime***

A **hydrograph** (see page 16) can be drawn for a particular point on a river (usually a gauging station). It shows a continuous record of fluctuating streamflow (discharge) measured in cumecs, and how a river or stream responds to precipitation.

A **regime** of a river shows average streamflow, usually measured over a minimum of 10 years. A number of factors influence the nature of a river's regime, as shown in Figure 4.

Rivers that have very interesting regimes to research are the Rhône, the Nile, the Yangtse-Kiang, the Ganges and the Colorado.

# Making sense of the hydrological cycle

The various stores in the drainage basin system are linked by flows. Each flow results from one of the processes defined below.

**condensation:** the process by which water vapour is converted into water.

**depression storage:** storage of water in hollows and holes in the ground surface to form puddles.

**evaporation:** the process by which water is changed to gas (water vapour) by molecular transfer.

**evapotranspiration:** loss of moisture to the atmosphere by the combined processes of evaporation and transpiration (via the stomata in leaves).

**groundwater store:** water held below the water table in aquifers.

**infiltration:** movement of water from rainfall or snowfall into the soil.

**interception:** the process by which a proportion of the precipitation input is caught and held by vegetation, especially where the water runs down trees as stemflow.

**overland flow:** the direct transfer of water to a stream channel by movement across the ground surface.

**saturated overland flow:** water gradually rises from the water table to the surface as the ground becomes saturated, and then seeps out across the ground surface.

**percolation:** the process by which water moves downwards through the rock (i.e. deeper movement below the water table, in the groundwater zone).

**precipitation:** deposition of water in either liquid form (rain) or solid form (ice/snow).

**runoff:** water that moves across the land's surface into streams, rather than being absorbed by the soil.

**streamflow (channel flow):** movement of water in a stream channel.

**throughflow:** downslope movement of water through the regolith.

## Review 2

Check that you understand all the basic terminology listed above. Then annotate Figure 5 to show how the hydrological processes are interlinked.

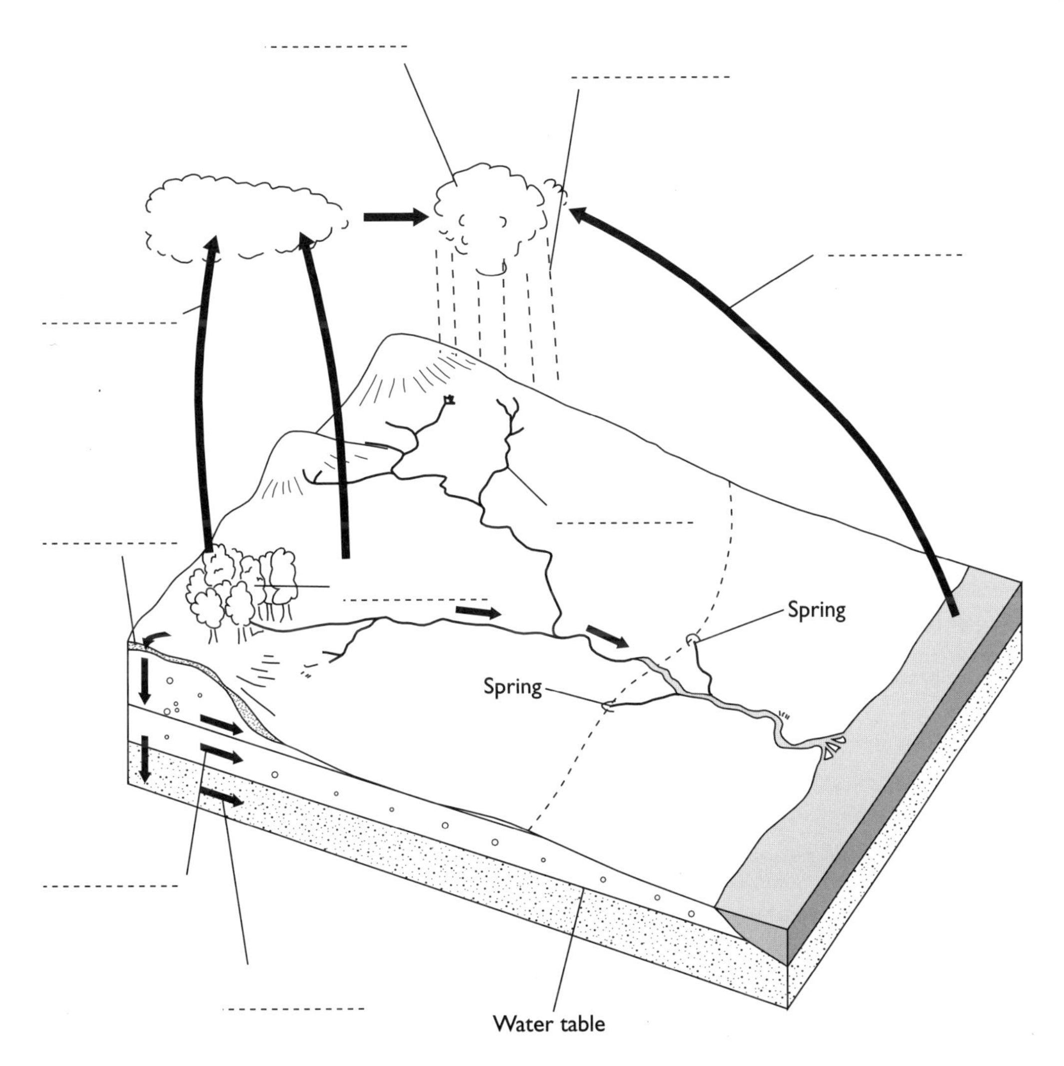


*Figure 5 Hydrological processes*

# Storm hydrographs

Exam questions requiring you to interpret storm hydrographs are very common, as they test your understanding of many aspects of hydrology. A storm hydrograph is a graph that shows how river discharge changes during a storm event, or a series of storm events. While many factors, such as river basin geology, affect the precise nature of the hydrograph, a standard terminology exists to describe key features of a storm hydrograph.

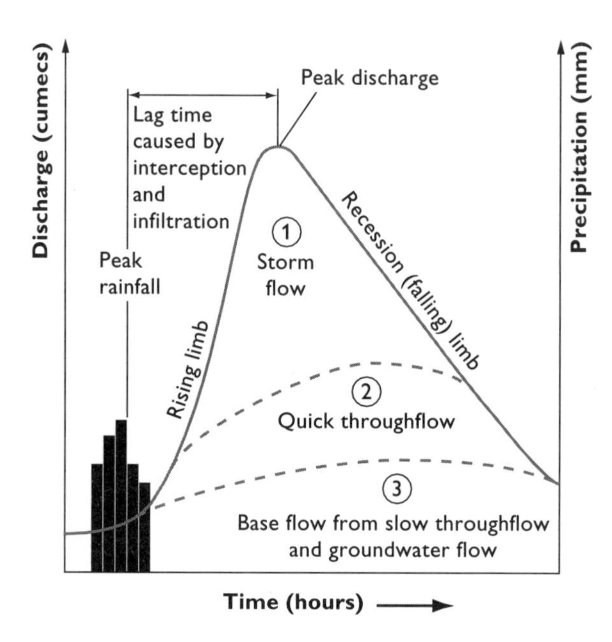


**Key**

1 Runoff from direct overland flow, or channel precipitation, reaches the stream very rapidly and leads to a rapid rise in discharge

2 Throughflow is slower, but nevertheless has an impact on the rising limb. It also contributes to the river when the storm flow has finished

3 Base flow may only arrive some 20–30 hours after the storm event, as the water has to seep into the system from deep below ground

***Figure 6 A storm hydrograph***

When *describing* a storm hydrograph, you need to include precise statements of amounts of precipitation, discharge, etc. over time and to use the terminology shown in Figure 6 above. When *explaining* the general form of a particular hydrograph, you need to think about hydrological processes and the sequence of how precipitation is passed through the various stores.

Numerous factors can affect the shape of a storm hydrograph.

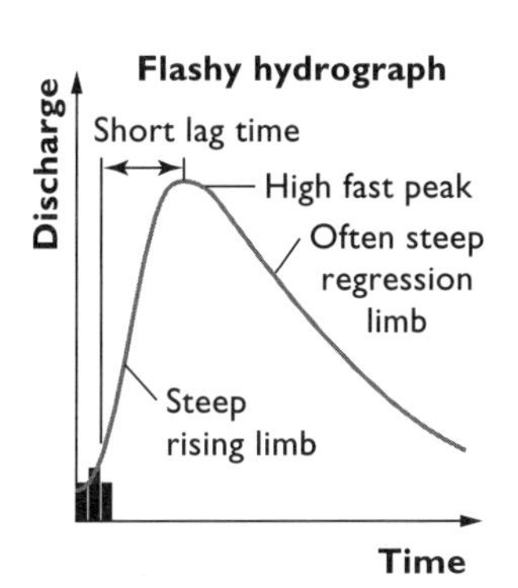


**Contributing factors:**
- impermeable rock type (e.g. clay)
- very intense storm with an early peak
- steep-sided river basin
- high drainage density
- small basin
- circular basin
- clay soil or soil with high antecedent moisture
- deforested catchment with short grass — often bare hillsides impermeable
- highly urbanised catchment — drains rush water to river

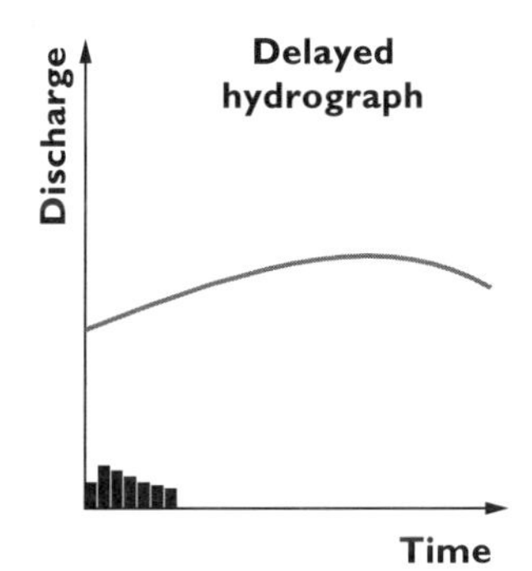


**Contributing factors:**
- permeable rock type with high percentage of base flow
- protracted, less intense storm
- more gentle-sided basin
- large basin
- elongated basin shape
- sandy soil with many pores
- low antecedent moisture
- forested catchment with high interception
- low levels of urbanisation
- possible impact of regulation from a dam

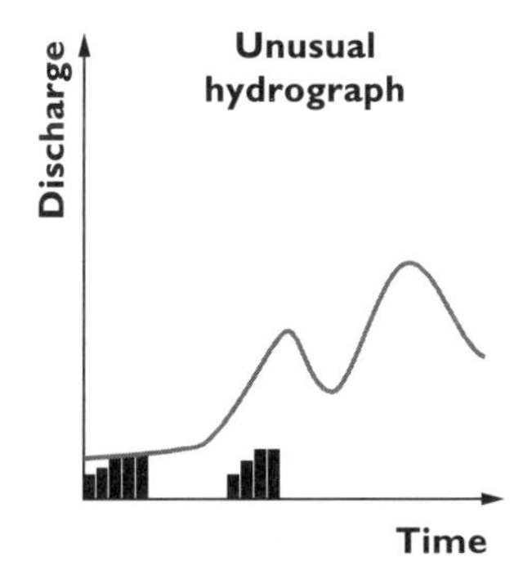


**Contributing factors:**
Just think it through — one solution is shown below:
- twin peaks related to storm events, in upper part of basin
- may be variable rock type and topography
- cloverleaf-shaped basin

***Figure 7 Three possible shapes of hydrograph and the likely factors leading to them***

## Review 3

(1) Using the information in Figure 7, explain the impact of the factors listed on the shape of each storm hydrograph.

(2) Figure 8 (page 18) shows two hydrographs for the *same* small stream draining an upland basin on *two separate occasions*.

(i) Compare the discharge curves shown.

(ii) Suggest reasons for any differences that you have noted in (i).

***Tip*** You must use precise figures and terminology, and for a comparison state 'whereas', 'on the other hand', etc. Do not just make two statements.

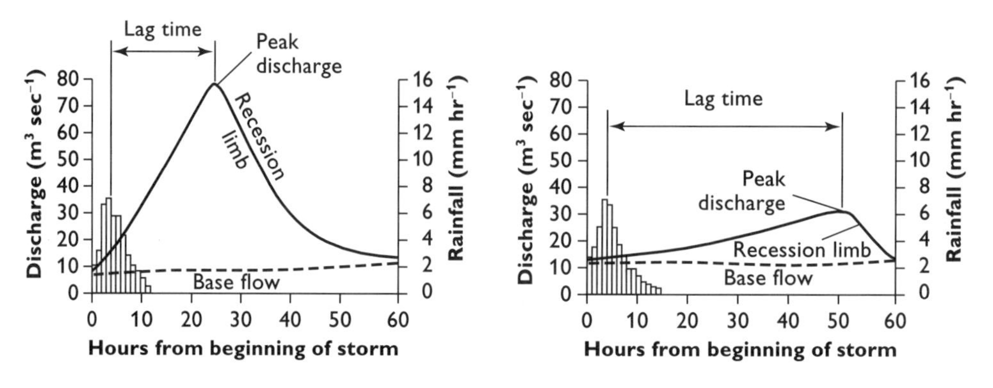


***Figure 8 Different hydrographs for the same stream***

***Tip*** You need to think of *short-term variables* and not *fixed* factors (see Figure 9), and to look at the rainfall event very carefully — it might be about the same amount, but...

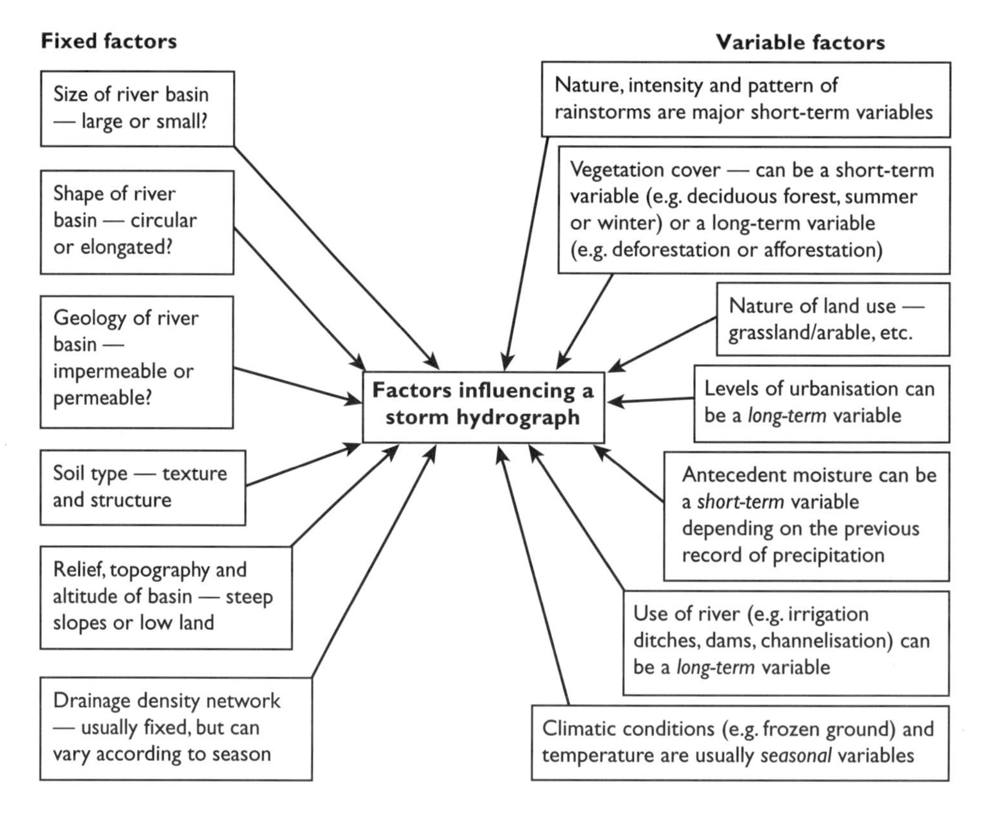


***Figure 9 The fixed and variable factors influencing the shape of a storm hydrograph***

# Processes and change in river environments

## The importance of river environments

River environments are highly valuable natural environments, but they are one of the most exploited for the following reasons:

- rivers provide a water supply for domestic and industrial users
- rivers are used as shipping routes and a means of waste disposal
- rivers are an important recreational resource (e.g. for fishing and water-sports)
- river channels and floodplains are mined for gravels and minerals such as tin
- rivers can be harnessed for hydroelectric power
- river floodplains are fertile and highly prized for agriculture
- river valleys and plains provide flat areas for route ways and building land, and are often densely settled

All this development requires management of the river, often using hard engineering defences, for example, to combat flooding. These management strategies may not be sustainable and can cause damage to aquatic environments within the river, and to the wetland ecosystems that naturally border rivers and streams. Damage results from drainage, pollution with fertilisers or channelisation. You are asked to look at rivers in a *holistic* way, including the ecosystems and environments of both the channel and the catchment.

## Long and cross profiles of a river: the basics

The **long profile** is a longitudinal section of a stream channel drawn from source to mouth. It therefore shows the **stream gradient**. A **theoretical** long profile shows a smooth concave curve. Here erosion, transport and deposition are in a state of equilibrium, with discharge and available energy in balance with available load. This theoretical profile is sometimes called a **graded** profile. It shows the impact of lakes and **knickpoints** (caused by localised bands of hard rock or rejuvenation), which form localised base levels (i.e. the lowest point to which a river can erode). Theory suggests that, over time, the knickpoint is eroded away, and aggradation fills the lakes to produce a smooth profile.

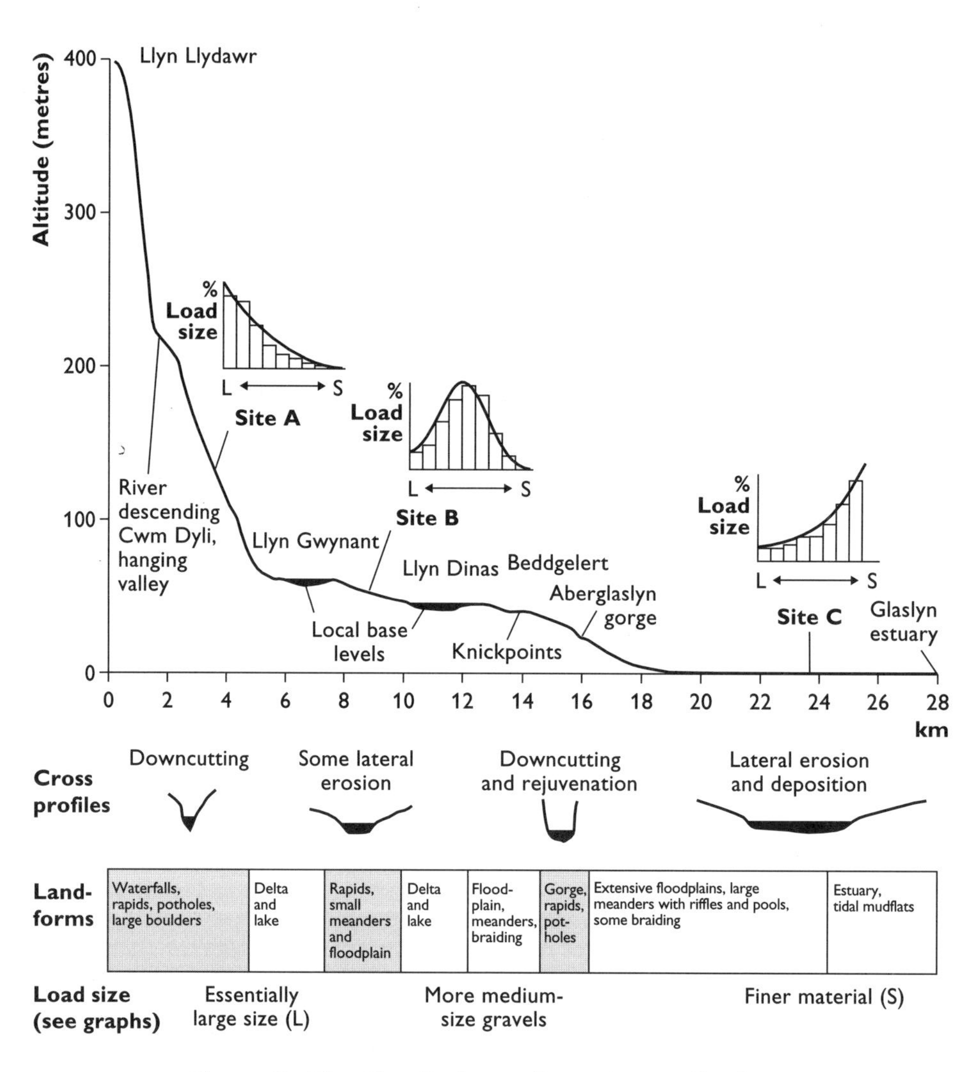


***Figure 10 The Aberglaslyn — the anatomy of a river***

Figure 10 shows how, in the case of the Aberglaslyn, rejuvenation has interrupted the standard long profile sequence to form the Aberglaslyn gorge. The diagram then goes on to show the classic features associated with each stretch of the Aberglaslyn, and how load size changes from source to mouth. It also summarises which processes are dominant.

**Cross profiles** can be drawn across a river valley. In the **highland parts** of river valleys, erosional energy leads to **vertical downcutting** and a V-shaped cross profile. The slopes above are sub-aerially weathered, with marked mass movement leading to supplies of debris falling into the river, but the valley widening cannot keep pace with the valley downcutting. As the river begins to erode **laterally** and downcutting is less apparent, a small floodplain develops, with slower sub-aerial processes leading to the occurrence of lower-angle slopes. In the **lower** course of the valley, as base level is reached, the main processes are lateral erosion and deposition to form an extensive floodplain with very gently sloping valley sides. The precise form of a valley cross-profile is influenced by the factors shown in Figure 11.

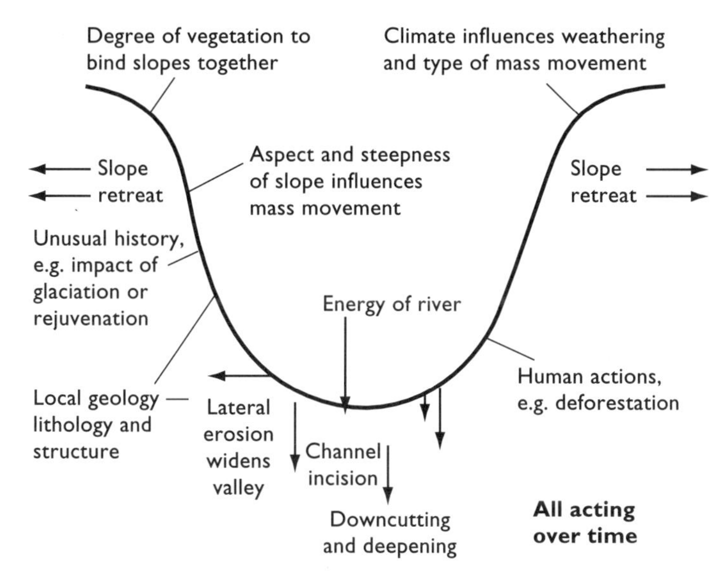


*Figure 11 Factors influencing river cross-profiles*

## Review 4

Choose a river about 12–15 km long that you can study. Use an OS map to draw long profiles and cross-profiles along its course. How do they compare with Figures 10 and 11?

# Understanding river channel variables

Figure 12 shows Bradshaw's model of how river channel characteristics change from the source (upper reaches) to the mouth. An alternative model is that developed by Schumm, which you will have a chance to analyse when you tackle exam question 1 on page 65.

A **model** is developed by measuring numerous real examples and represents the **theoretical** situation. When you survey a stream or river in the field and make measurements of the variables shown, you relate your findings to the model and see whether they are similar or not. You then try to seek explanations for any **anomalies**.

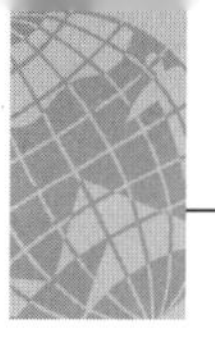

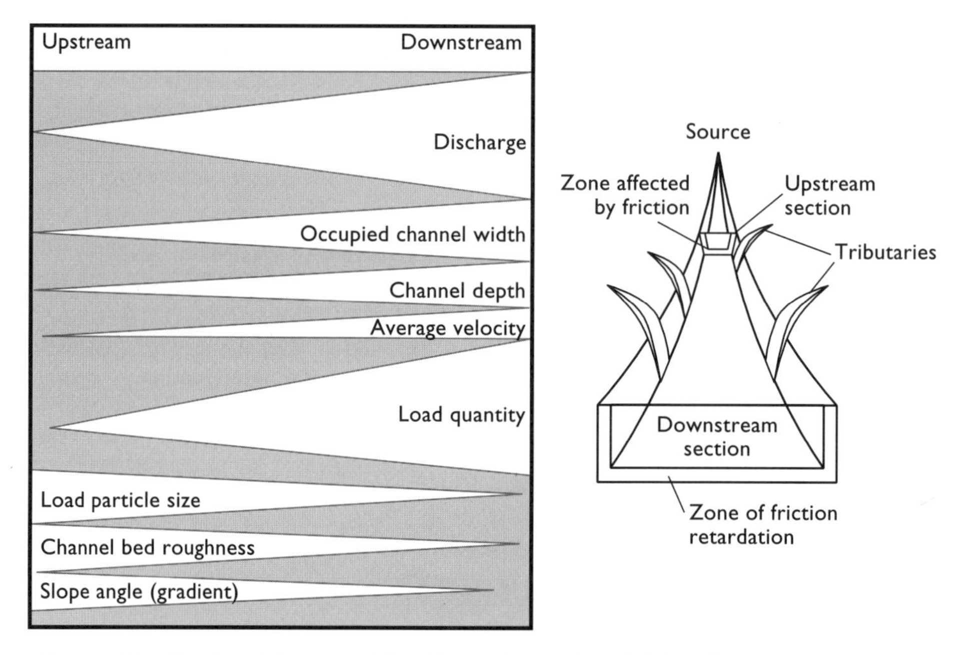


***Figure 12 The Bradshaw model of how channel variables change downstream***

The elements of the Bradshaw model can be explained as follows:

- **Occupied channel width** is the distance across the actual channel, measured at the water surface. It can change according to seasons. On average, it increases downstream as tributaries join the river.
- **Channel depth** is the height from the water surface to the stream bed. In natural channels, as the bed is uneven because of large boulders, the mean of several measurements is used. Again, as you would expect, depth increases downstream as the river becomes larger, *but* there are often localised variations: for example, those associated with **riffles** and **pools**. A line connecting places of greatest depth is called the **thalweg**.
- **Cross-sectional area** (*A*) is closely linked to width and depth and can be calculated in a number of ways. It is the product of width and mean depth, $A = WD$. A cross-section can be either drawn by hand or plotted using a computer program.
- **Wetted perimeter** (*P*) is the length of the channel margin (bed and banks) around the cross-section which is in contact with the water, and it can be measured either in the field or from your graph. It is a vital measurement when you are investigating channel efficiency, which is calculated as $A/P$ (cross-sectional area divided by wetted perimeter). The ratio is known as the **hydraulic radius** (HR) — the higher the number, the more efficient the channel.
- **Average velocity** (*V*) is the speed of water flow (i.e. the distance travelled per unit of time) and can be measured using timed missiles (tangerines for a large river) or a flow vane. It is usually measured in metres per second. When you look at the

Bradshaw model, you can see that *V* steadily increases downstream. Students find this very hard to understand when they can see that the **gradient** (slope of long profile) is decreasing. The explanation is given in the three-dimensional diagram in Figure 12. This shows that downstream, because the river is more **efficient** (higher HR) with proportionally less contact with its bed and banks, and the **channel bed roughness** has decreased because of fewer stones, there is less **external friction** to slow the water down. There is also less turbulence. The lower amount of friction counteracts the reduced gradient.

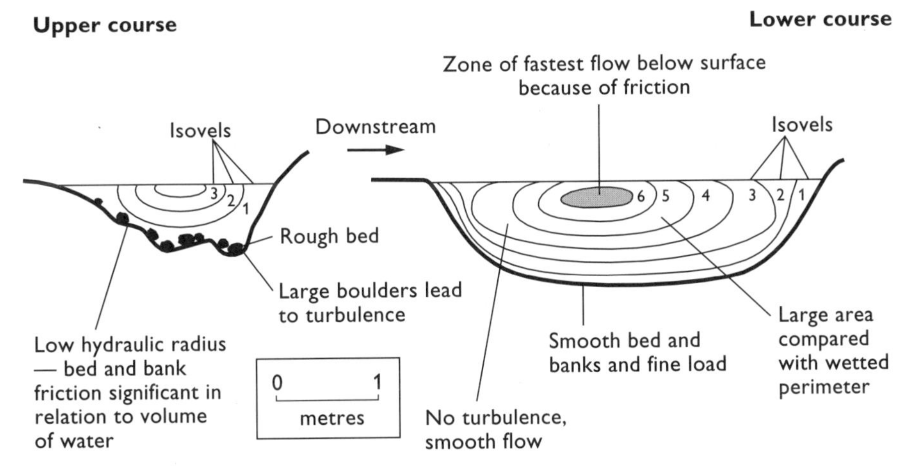


***Figure 13 How cross-sections and velocity differ downstream***

- **Discharge** (*Q*) is the volume of water that passes through a cross-section (*A*) per unit of time. It is usually measured in cubic metres per second.

  $$Q = \text{average velocity } (V) \times \text{cross-sectional area } (A)$$

  In practice, the average velocity can be calculated by multiplying the mean surface velocity by 0.55 — if only floats are used. For a flow vane, the average velocity is formed around 0.4 of the way up from the river bed (average velocity = 0.4 depth). As both cross-sectional area and velocity increase, the *model* shows a very large increase in the amount of discharge from source to mouth.
- **Load particle size** (*L*) (i.e. the **calibre**) decreases considerably downstream (**comminution**) and the angular pebbles become more rounded, largely as a result of the process of **attrition**. The mean size and shape of sample groups of individual pebbles can be measured using a stone board.
- **Load quantity** (i.e. the **capacity**, or total load of all sizes) increases as the discharge increases — they are **directly proportional** in their relationship.

## Review 5

An **isovel** is a line of equal water velocity. Obtain your own data and try drawing them accurately. Compare the cross-section of a meander with that of a straight section.

# Drainage basins and sediment yields

Rivers and streams obtain their load from two sources — 90% from weathering and mass movement of slopes within the catchment, and the rest from the river's own channel bed and banks. River channels are therefore seen as **transfer systems**.

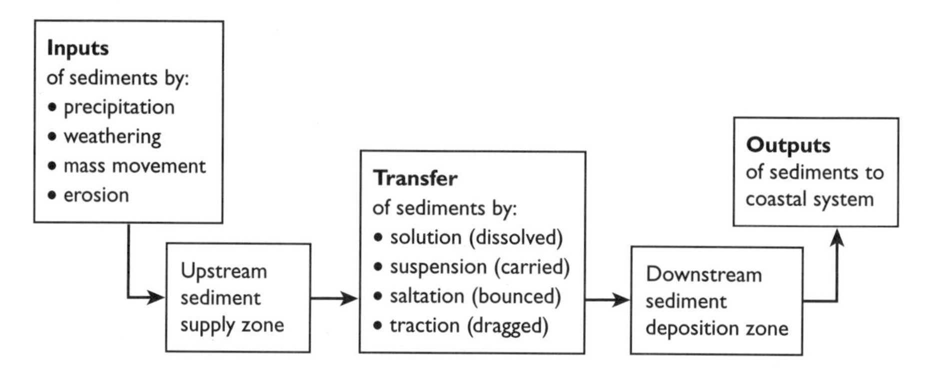


***Figure 14 The transfer of sediments***

The factors that influence how much sediment (the yield) is input to the river system include many of those that influence runoff (see Table 2, on page 11). The world's major rivers show considerable variation in sediment yield because of these factors. The largest load carrier of all is the Yellow River (Hwang He) in China, where the geology (a fine material called **loess**) is easily transported. Where people interfere in river systems, the effects on sediment yield can be dramatic. Deforestation can increase yields, as in the Ganges headwaters, while dams can decrease the supply, as on the Nile. What happens in river systems can have major impacts on coastal systems, as seen in West Africa (see page 40), and along the Californian coast where the San Gabriel no longer supplies sediment to the beaches.

# Erosion and entrainment

To become part of the river's load, a particle must first become dislodged by erosion, then picked up and set in motion. This process is called **entrainment**.

Erosion can be:

- vertical (**corrasion** and **abrasion**), which increases channel **depth**
- lateral (**hydraulic action**), which increases channel **width**
- headward (**spring sapping**), which increases channel **length**

The three Cs are important terms to know when writing about a river's load. These are:

- **capacity** — the total sediment load of a river at a particular time or location
- **calibre** — the size of a particular pebble or particle

- **competence** — the size of the largest sediment particle that can be carried by that river at a particular time or location

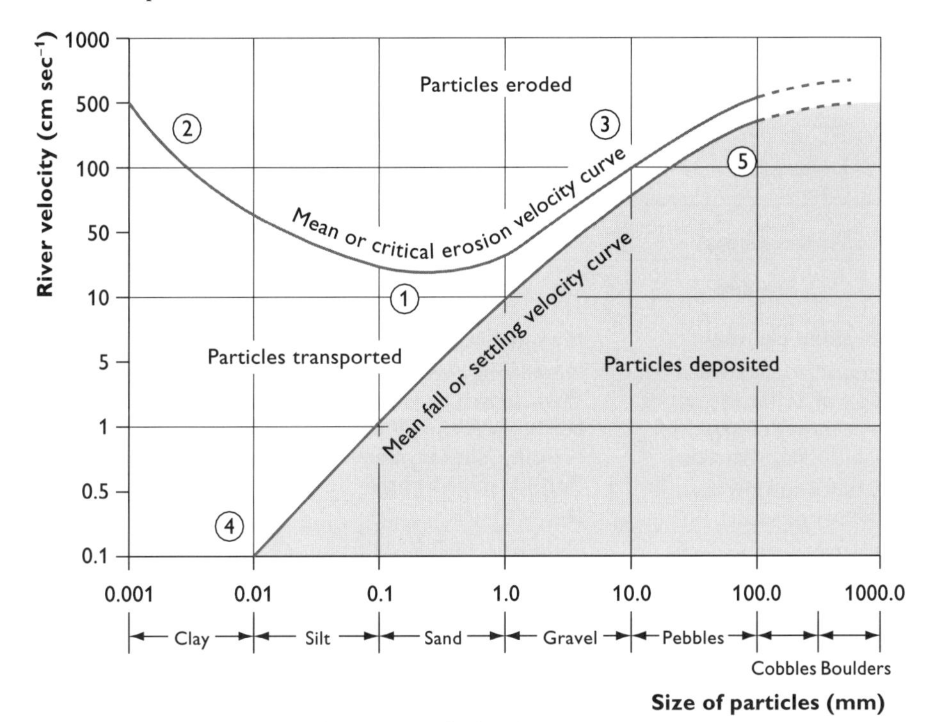


***Figure 15 Hjulström curve***

For a particle to be picked up, the **critical erosion velocity** has to be reached. Figure 15 shows that sand can be transported at lowezr velocities than either finer or coarser particles. At point 1, particles of about 0.2 mm diameter can be picked up by a velocity of 20 cm $sec^{-1}$. Finer clays at point 2 and larger pebbles at point 3 require higher velocities to be dislodged. This is because clay particles, though lighter, are cohesive and so stick to the river bed.

The velocity required to keep particles in **suspension** is less than that needed to pick them up. In fact, for very fine clay particles at point 4 in Figure 15, the velocity is almost nil. So once picked up in turbulent tributaries, these particles may be carried long distances down the river system.

The fall or **settling velocity** curve shows the velocities at which particles can no longer be carried and so are deposited. Large boulders shown at point 5 in Figure 15 are deposited first, having been carried only short distances, and are therefore often seen in the bedload of mountain streams. **Saltation** moves gravels, silt and sand to infill valley floors, while the finer clay and silt in suspension reaches tidal mudflats. Deposition increases downstream, and depositional landforms occur in low-energy situations, such as at water margins, at changes in gradient or at river mouths.

# River channels

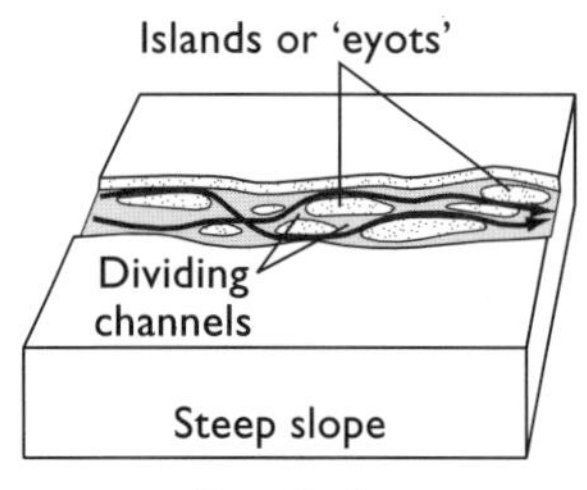


**Braided**

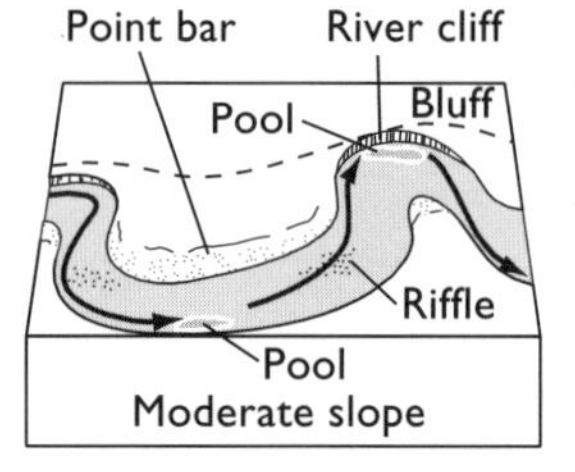


**Meandering**

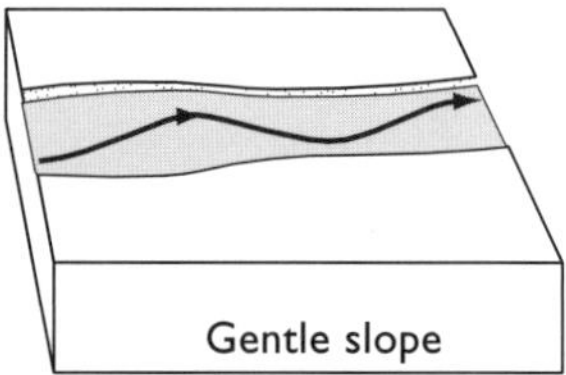


**Straight**

**Braided channel**

Braiding occurs when the flow of a river threads its way through a series of interlocking channels.

Braided streams are associated with:

- steep river gradients
- wide valleys floored by coarse sands and gravels

Braided stream regimes frequently have very variable flows, with large floods that destroy and rework the island material.

Braided channels are far less stable and frequently change course.

Occasionally, vegetation colonises the islands.

**Meandering channel**

Meandering rivers are characterised by loops and bends of the river. If the sinuosity index is greater than 1.5, meanders are said to occur.

Meanders result from helical flow caused by the spiralling of a core of maximum velocity. They migrate to form a floodplain bounded by bluffs. Meanders show changes in depth known as riffles and pools.

Meanders are associated with moderate gradients and areas of finer (silt) bank sediments.

**Straight channel**

Straight channels are associated with gentle gradients in areas of steep, even vertical, banks that constrain the river flow.

Straight stretches can be the result of channelisation.

***Figure 16 Rivers in plan — river channel traces***

The change from one form of channel shape to another can occur in two ways:

- an external change in energy, such as results from human interference (e.g. building a reservoir)
- a slow natural internal change (e.g. readjustment after a storm event)

# The river environment as an ecosystem

The value of a wetland habitat depends on many factors, such as age, size, depth of water, nutrient levels and water chemistry. Interactions between these factors and ecological processes operating over time are responsible for the range of plant and animal communities present in a particular habitat. Fresh water is one of the richest wildlife environments. It has a high total number of species with a high diversity, and high **primary productivity**. Some examples include:

1. **wetlands** — along the channel margins, marshy areas fed by runoff and periodic flooding
2. **oxbow lakes** — abandoned meander arcs that can become ponds or be periodically flushed out by the river in flood
3. **ponds** — depressions in the floodplain filled by water from runoff or surface detention
4. **wet woodland** — damp areas of deciduous trees holding water and providing shade
5. **washlands** — areas of the floodplain that are regularly covered in standing water in flood conditions, usually deliberately
6. **chutes** — channels of more rapid water flow often between point bars and slipoff slopes
7. **riffles and pools** — areas of shallower and deeper water (respectively) in river channels
8. **point bars** — depositional and island features which change seasonally

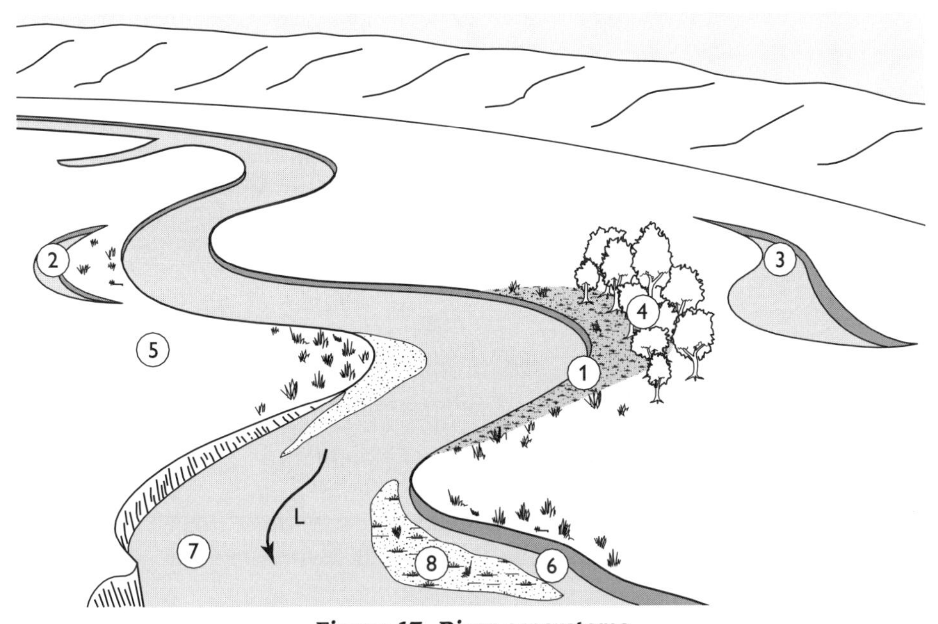


***Figure 17 River ecosystems***

The **structure** of an ecosystem is how its parts are arranged. In a pond a number of physical factors, such as light, temperature, availability of oxygen and acidity determine how the plants and animals are arranged. In Figure 18 you can see three main zones of life in the profile of a pond:

1. the **surface film** — a region of abundant oxygen and maximum light
2. the **vegetation zone**, which includes the margins of the pond. It is a zone of **high primary productivity** and supports a wide range of herbivores, which in turn

provide food for numerous carnivores, such as pond snails, leeches and diving beetles

3 the **mud zone**, with a high organic content in the form of detritus. It provides a source of nutrition for microorganisms and serves as a home for burrowing worms. Over time, this mud zone will increase in size and this can trigger off a **successional sequence**

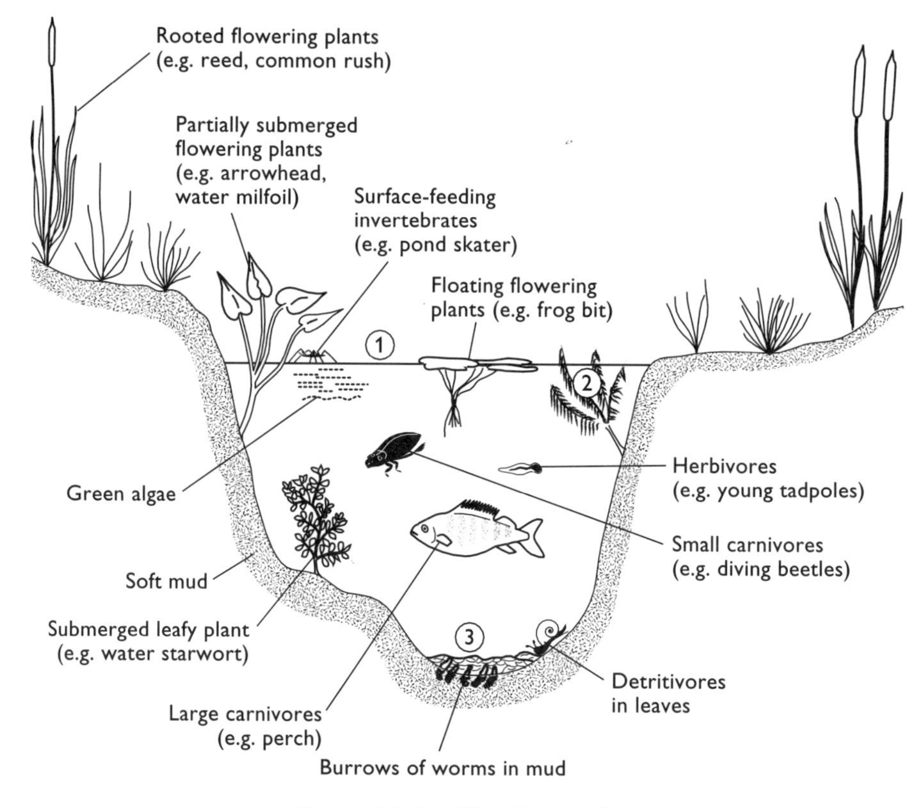


***Figure 18 Profile of a pond***

The **structure** of an ecosystem can also be thought of in terms of **trophic** (or feeding) levels, with each successive level being supported by the one below it.

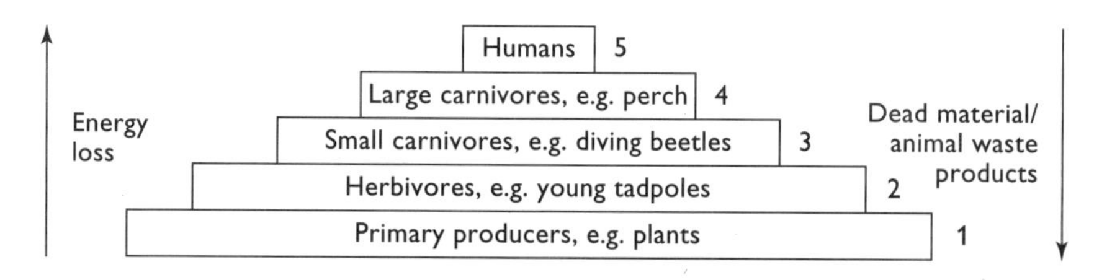


***Figure 19 Trophic pyramid in a pond***

The **functioning** of an ecosystem is how it works — how **energy flow** occurs and how **nutrients** are cycled. The movement or flow of energy from plants to animals to decomposers can be represented in its simplest form as a **food chain**. Each stage in the chain where energy is exchanged is known as a **trophic level**. Green plants absorb light energy from the sun by **photosynthesis**. The rate of energy production is referred to as **primary productivity** and this is very high in swamps and marshes. At each trophic level, energy is lost in the form of heat, as a result of respiration.

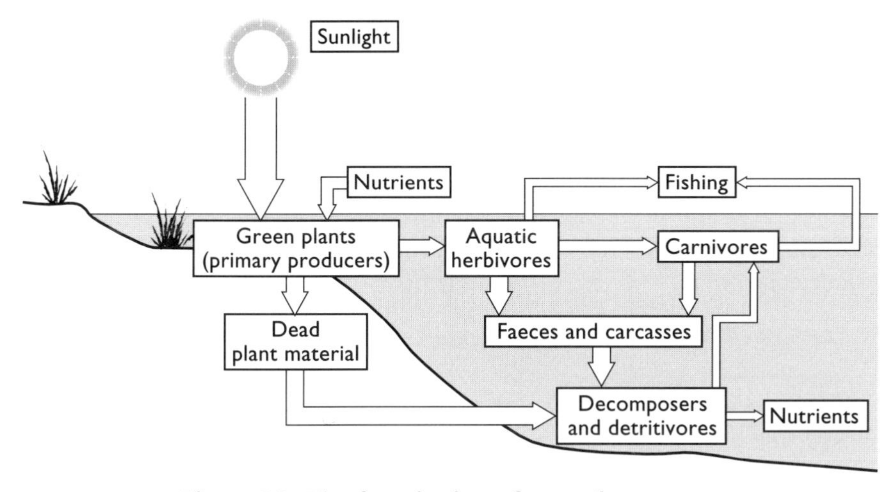


***Figure 20 The functioning of a pond ecosystem***

***Tip*** You need to adapt these generic diagrams into specific case study ones — for example, a named pond or a particular oxbow lake or wetland — and refer to appropriate flora and fauna.

## Review 6

(1) Define 'primary productivity' and 'decomposers'.
(2) Suggest two ways in which humans can affect the lake or pond ecosystem.
(3) Explain how energy flows occur in the lake or pond ecosystem.

# Environment–people interactions in river environments

## Changing river environments

Rivers are said to be in a state of **dynamic equilibrium** — that is, in a state of balance which is constantly adjusting to *routinely* changing conditions. However, a severe sudden storm or dramatic snowmelt, or a prolonged change in inputs or outputs, can lead to dramatic changes in discharge or sediment levels, thus upsetting this state of dynamic equilibrium.

Two examples are shown in Figure 21.

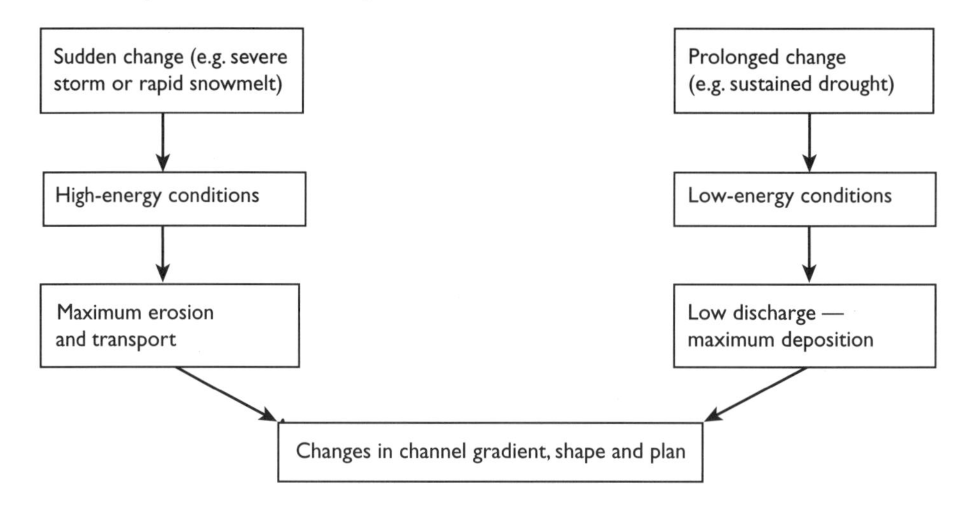


**Lynton, Devon**

In August 1952, 230 mm of rain fell in a freak storm lasting 14 hours. As the ground was saturated, the surface runoff caused a flash flood with a discharge level far higher than that of the River Thames. The size of the bed load was enormous, as over 100 000 tonnes of boulders were left in the main street. Rocks up to 10 $m^3$ were moved by traction.

**Southern England**

In July–August 1976, blisteringly hot conditions (a blocking summer anticyclone), constant 25–31°C temperatures and no rainfall led to extreme drought conditions. Many rivers dried up or diminished to a trickle, leading to the deposition of large quantities of load.

***Figure 21 The effect of extreme weather on river environments***

While many farming practices rely on seasonal flooding or silt deposition, few river environments can cope with severe, dramatic or sudden changes in discharge and sedimentation. The need to manage rivers so that these dramatic changes have minimal impact is further increased because many river environments are very densely populated.

Just as the dynamic equilibrium of river systems can be upset, so can that of coastal systems. A severe coastal storm redistributes beach deposits, breaches marine bars, and causes dramatic cliff and shore erosion. In both cases, it may take considerable time for the system to readjust.

While sediment from the river can benefit riverside farmers, too much can cause problems. It can:

- limit the channel capacity and increase the risk of flooding
- clog up irrigation channels
- infill lakes and pile up behind dams, thus decreasing their efficiency
- block shipping channels
- choke fluvial ecosystems by blocking out the sunlight in photosynthesis
- block fish gills
- discolour and pollute the water supplies, as sediments may be contaminated

There are a number of river 'problems' that humans are concerned to manage, the most common of which is flooding. These problems are frequently interconnected, and some solutions are appropriate for several problems.

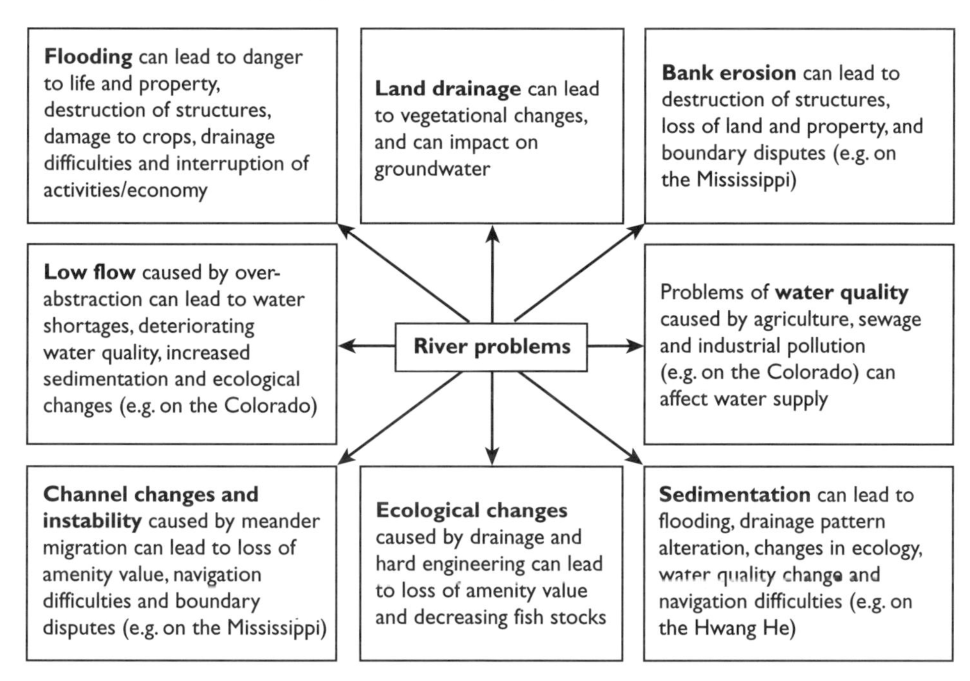


***Figure 22 River problems***

# A case study of flooding

Flooding is a problem that nearly all river authorities have to manage. It is expensive because it damages property, and it can also cause loss of life. What is a usual river process can become a hazard when a number of physical conditions occur, and human actions can both directly and indirectly increase **risk** from flooding. As more and more people have chosen to settle beside rivers, they have increased not only the risk of flooding, but also the need for management strategies that make their homes secure.

Water managers want to know how often floods can be expected (**frequency**) and how severe each occurrence will be (**magnitude**). They can assess risk based on prediction from discharge data records in past years. They calculate the **recurrence level** (the interval at which a particular level of flooding is likely to occur). Managers can then use this information to find out the area at risk and depth of flooding that would be caused, and recommend zoning regulations. The big problem is that increased urbanisation and changing agricultural use (more cropping) in catchments is actually increasing the frequency and magnitude of the discharge (see Figure 23).

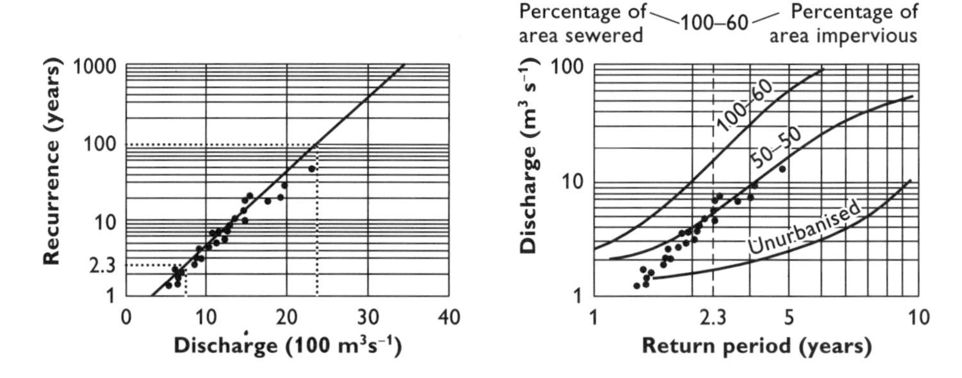


***Figure 23 Contrasting flood frequency curves***

## Review 7

(1) What relationship is shown in the first graph?

(2) What is the effect of more sewers and concrete on the second graph?

***Tip*** You will find flooding a core concept in the natural hazards option in Unit 5. It is a good idea to research flooding on a variety of rivers. Try the following websites:

**www.environment–agency.gov.uk**
**www.nwl.ac.uk**
**www.rivernet.org**

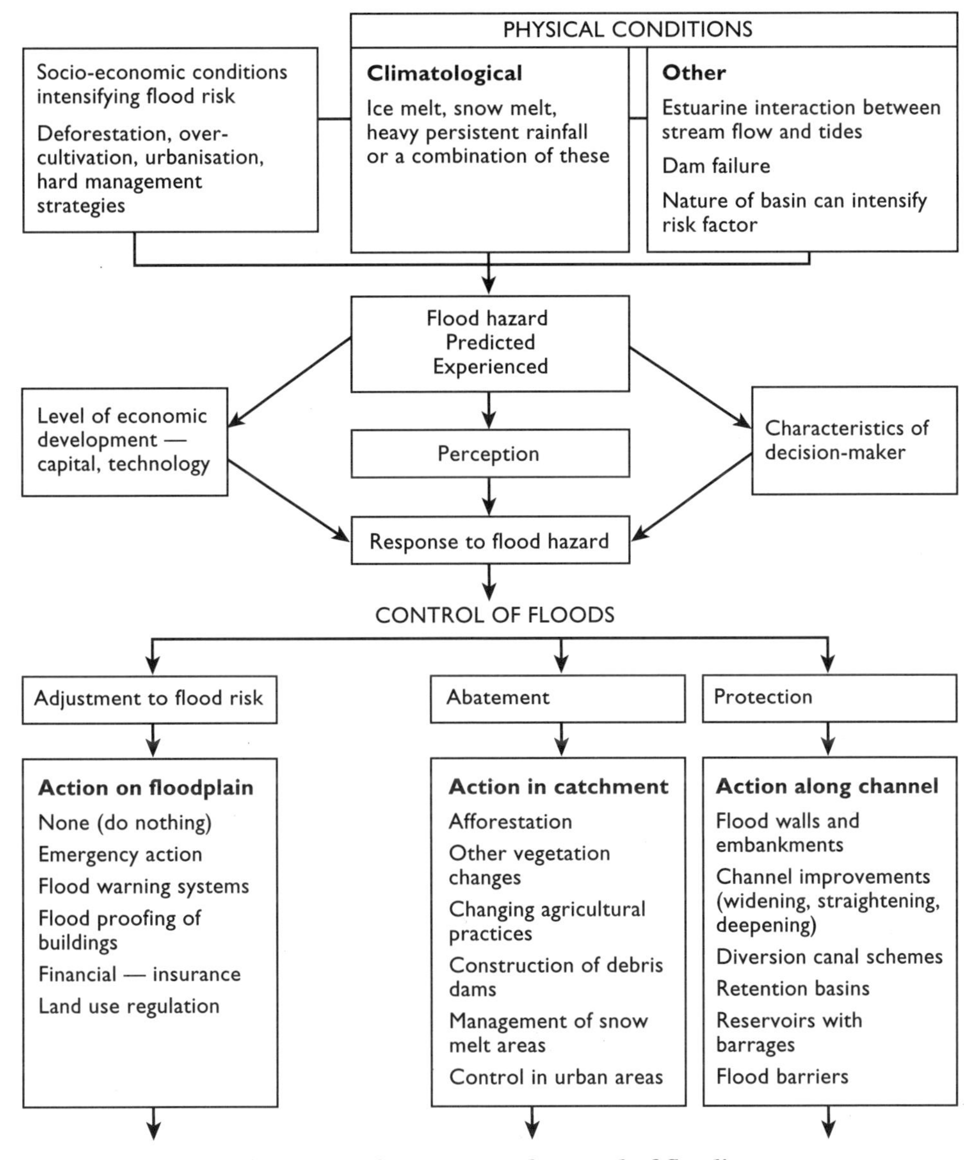


***Figure 24 The causes and control of flooding***

## Review 8

As can be seen in Figure 24, there are three responses that can be made to flooding. Identify the characteristics of adjustment, abatement and protection.

***Tip*** Consider the period before, during or after floods, short or long term, and whether responses are aimed at causes or consequences. In each case, the benefits have to be assessed against the economic and environmental costs.

# A detailed assessment of two management strategies

This section looks at the impacts of dams and channelisation — two very widely used management strategies.

## Dams — a blessing or a curse?

Dams are built for a number of purposes — water storage, hydroelectric power (HEP), flood control or regulation of river flow for navigation. Many **mega dams** (very large dams) have to be multipurpose in order to justify their multimillion pound costs. Dams have a major impact on the regime of a river as they enable discharge to be regulated. The lake upstream from the dam (storage reservoir) provides a new store for river sediment. Water emerging from a dam therefore has a new regime, a changed discharge and a different sediment load.

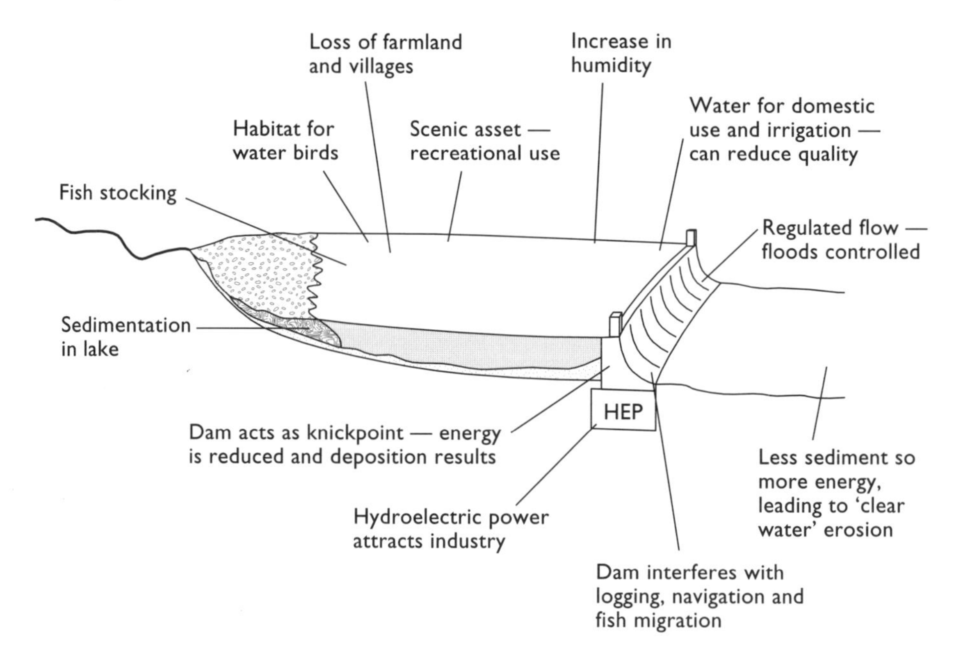


***Figure 25 The impact of dam construction***

Figure 25 summarises some of the environmental and socioeconomic impacts of dams. Clearly, the larger the dam, the more dramatic its impact on the environment and ecology of the area. While the development of mega dams seems to make economic sense, there are many disadvantages. HEP is considered to be clean and renewable, but there is an enormous environmental downside.

## Review 9

You will need to have learnt a range of case studies to look at the costs and benefits of these mega dams — both single examples, such as Aswan, and a series of dams along a river, such as on the Colorado. (See www.dams.org)

## Channelisation — the number one strategy?

Channelisation is a deliberate attempt to alter the natural geometry of a water course. It is therefore an example of **hard engineering**. It is a very common strategy and is used, for example, along the Mississippi River. (See www.mvs.usace.army.mil/eng/river)

The reasons for using channelisation can include:

- increasing the capacity of a channel to prevent flooding
- providing a straighter and deeper channel for navigation
- preventing bed or bank erosion
- reclamation of wetlands by lowering the water table
- straightening rivers to make farm land more manageable, or to allow bridges or highways to be built more easily

Channelisation is carried out as follows:

- **Resectioning** involves enlarging the cross-section of the channel by **deepening** or **widening** the river to increase its hydraulic efficiency, and to allow a larger discharge to be contained within the channel. **Dredging** is used to remove surplus sediment from the river bed and **vegetation clearance** is used to remove it from the bed and banks.
- **Realignment** or **straightening** can be done by means of artificial cutoffs. The aim is to increase the long profile gradient so that there is an increase in velocity and flood waters can be moved more quickly. Straightening can improve navigation. Downcutting occurs upstream of the realigned channel and deposition downstream.
- To reinforce channelisation, banks can be protected with concrete blocks, steel **revetments** or **gabions** to stop meander development. **Wing dykes** and **training walls** are built out from the river banks to focus the river current in the centre of the river. Often a **resectioned** length of river in urban areas is **lined** with concrete. This reduces friction and increases channel efficiency.
- In problematic sites, **containment** can occur using **culverts** to guide streams underground through urban areas.

The impacts of channelisation are complex. Usually the desired aims of flood management or improved navigation are achieved only at a cost to the river environment and ecology, and sometimes these aims even conflict with each other. Channelisation has impacts on river channel processes, hydrology and water quality as well as on river ecology and land drainage.

Figure 26 aims to summarise these impacts. Figure 27 ranks the types of channel modification in terms of their ecological damage.

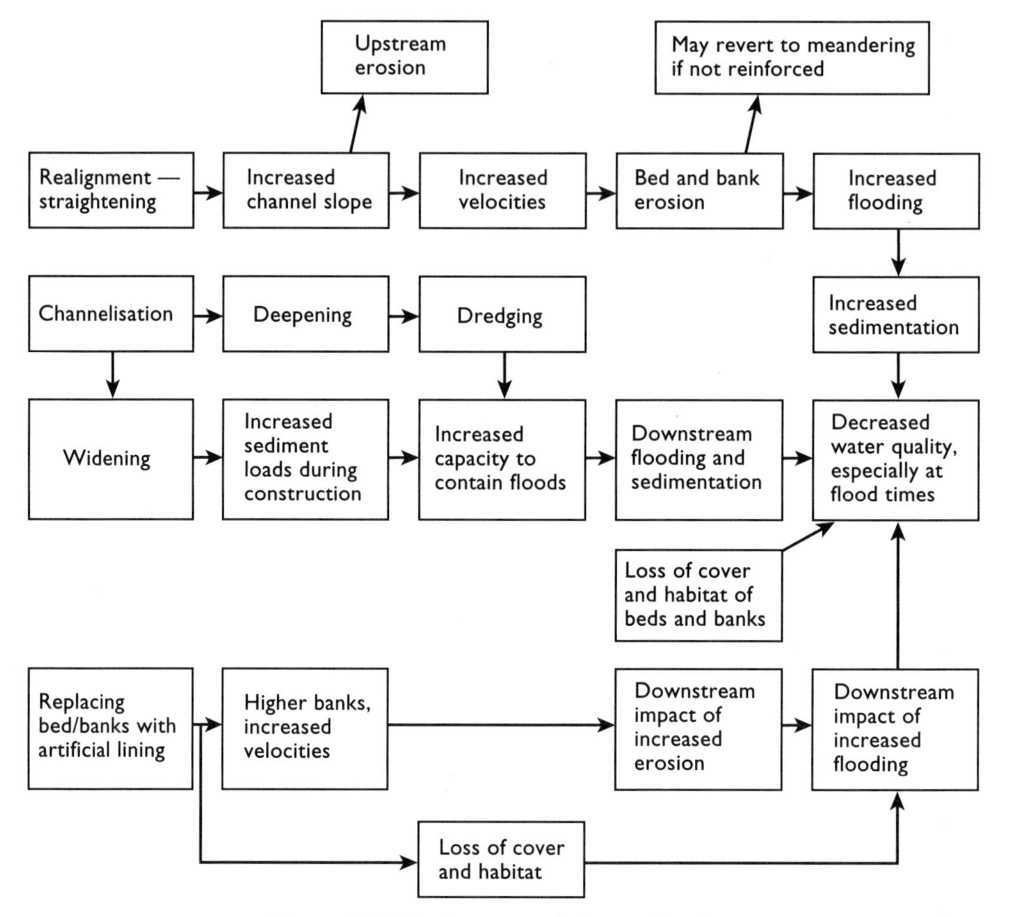


***Figure 26 The impacts of channelisation***

# River management and the future

Channelisation and dam building are examples of hard engineering strategies developed to alleviate problems of flooding and river channel adjustment. There are numerous environmental and economic costs that have to be weighed against the benefits. These hard engineering options are not considered to be sustainable. Figure 27 shows how the various hard engineering options compare in terms of their impact on the river environment and ecology.

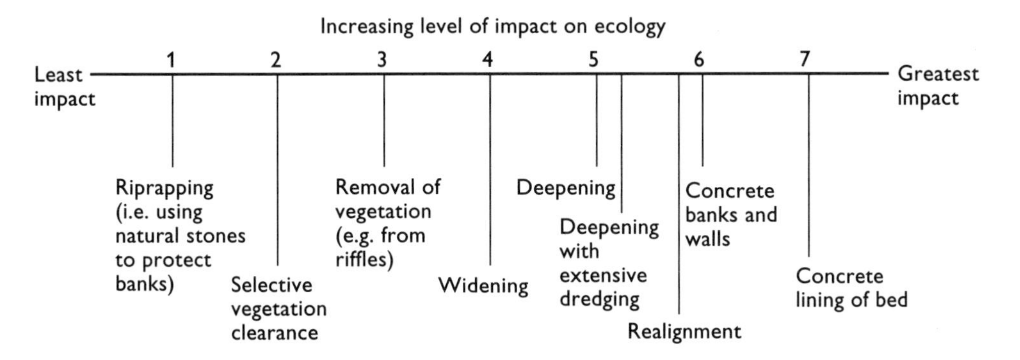


***Figure 27 The ecological impacts of management strategies on river systems***

In this section, we look at possible alternatives to hard engineering. A number of revised river management strategies are sympathetic to maintaining a quality river environment, in which engineers work with the river. These are holistic strategies that treat the river both as a dynamic geomorphological system and as an ecosystem. They are thus sustainable.

In some cases, river environments that have been destroyed by hard engineering are actually being **restored**. The same US engineers who once straightened the Kissimmee River in Florida, to help drain the land, are now putting the meanders back to conserve its environment. (See www.sfwmd.gov/orb/erd/krr)

River restoration is an expensive and complex process. Projects are under way in the USA, Germany, Denmark and the UK. They range from small catchments to short stretches of larger rivers such as the Rhine. (See www.qest.demon.co.uk)

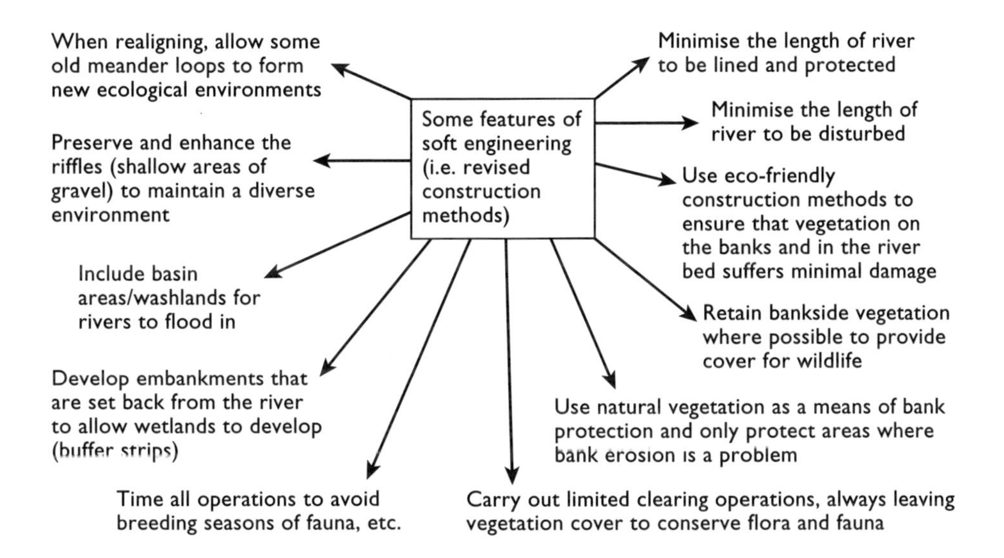


***Figure 28 Features of soft engineering***

To manage rivers in the future, managers are looking at a number of options, all of which should cut down on the use of hard engineering strategies.

- **Integrated catchment management** involves developing an overall plan (in the UK, the Environment Agency is doing this). Managers look at a problem such as flooding and evaluate a range of options — including afforestation schemes that manage the longer-term causes of flooding as opposed to tackling the short-term effects. Hard engineering schemes will still be used in key areas — for instance, where the river environment is highly urbanised.
- **River corridors** or **stream ways** can be created. These corridors can be set aside only where floodplain development has not occurred, as they give rivers room to move and make the assumption that they are meant to flood.
- **Sustainable use of wetlands**, which occur alongside river areas and are noteworthy for their high ecological value and high productivity. Hard engineering has in the past drained wetlands, usually for commercial agriculture (e.g. the Halvergate Marshes). These wetlands can be natural storage reservoirs against flooding and buffer zones against pollution. Additionally, modern thinking sees wetlands as worthy of conservation, as nature reserves or as areas of farming. In environmentally sensitive areas, farmers are actually paid to farm in a way that conserves these wetlands. Wetlands are also important for recreation and tourism. Some badly damaged wetlands, such as the Danube Delta, are now being restored.

River managers of the future will have a very hard task as a result of physical and human factors that are changing.

- **Global warming** (which you will study in depth in the global challenge unit in A2) could provide river managers with a huge challenge. Figure 29 takes you back to the hydrological cycle to assess this challenge.

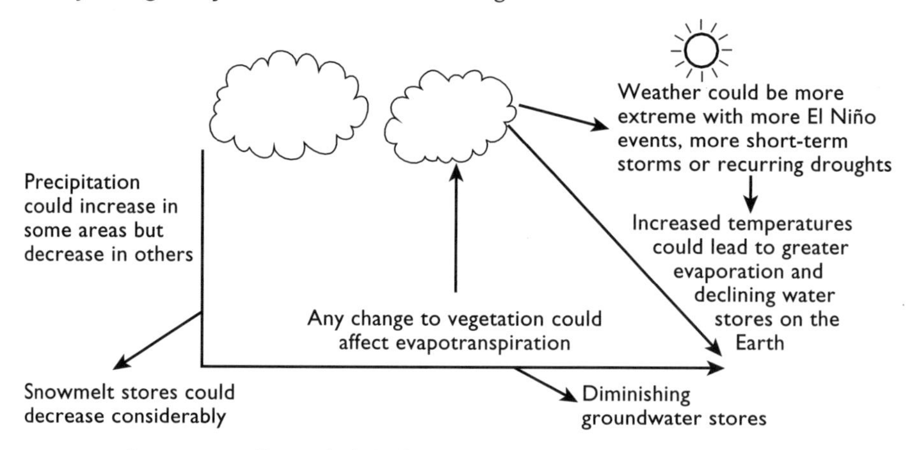


***Figure 29 Effect of global warming on river management***

- The human dimension results from the **exponential growth of population**, and its food, water and housing needs. While sustainable strategies for managing the river environment are vital, sustainable strategies for managing people's use of water are the key.

# Introducing coastal environments and systems

## The importance of coastal environments

Coastal environments play an important part in the physical and human geography of most countries. A number of key ideas and issues are involved:

- the 'coastal zone' is an **interface** between the sea and the land, where marine and terrestrial processes combine to produce a variety of changing landforms
- coasts are hazardous because rapid erosion and flooding can threaten lives and property
- changes to coastlines are both short term (e.g. in storms) and longer term (rising sea levels)
- coastal ecosystems, such as sand dunes and coral reefs, are coming under increasing pressure from human activities
- coasts encourage human settlement and economic activity — over half of the world's population and two-thirds of its largest cities are within 60 km of the sea
- the coastline is a 'frontier' where there is competition for, and conflict over, natural resources
- there is a need to manage existing coastal environments more effectively
- human intervention can upset natural systems and might lead to unexpected impacts
- in the future, increasing development and the impacts of global warming will demand yet more careful, and where possible **sustainable**, management strategies

## Understanding the coastal system

The coastal system is an **open system**. River deltas and estuaries are significant features of coastlines. They deliver large quantities of fresh water and sediment into the coastal system from the fluvial system. Sub-aerial processes, waves and tides add to these **inputs**. **Transfers**, such as longshore drift, move sea water and material through and out of the coastal system, and beaches are the major sediment **stores** (it is estimated that 90% of beach material is derived from rivers). Where there is a balance between these various inputs and **outputs**, the system is said to be in **dynamic equilibrium**. Figure 30 shows the main features of this system.

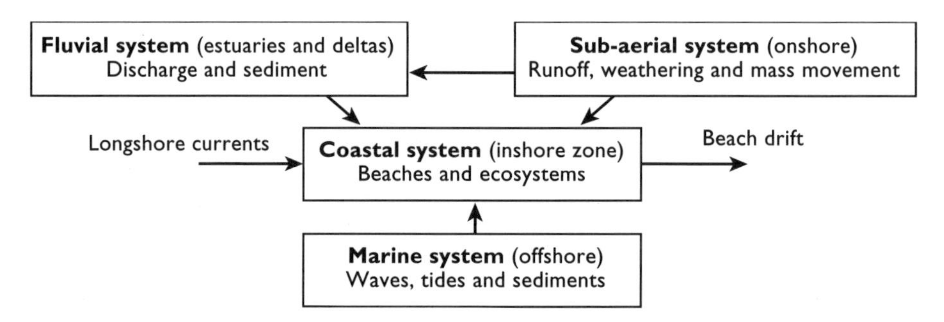


***Figure 30 The main components of the coastal system***

# The impacts of human intervention

Human intervention in the fluvial system or coastal system may have damaging as well as beneficial effects. This is clearly seen in west Africa, where the building of a dam on the River Volta at Akosombo has not only caused changes to the river environment, but also had an effect on the coastlines of Ghana and neighbouring countries. This shows how closely river and coastal systems are linked. Figure 31 shows the effects of this intervention. Another example is found in Question 2 (page 69).

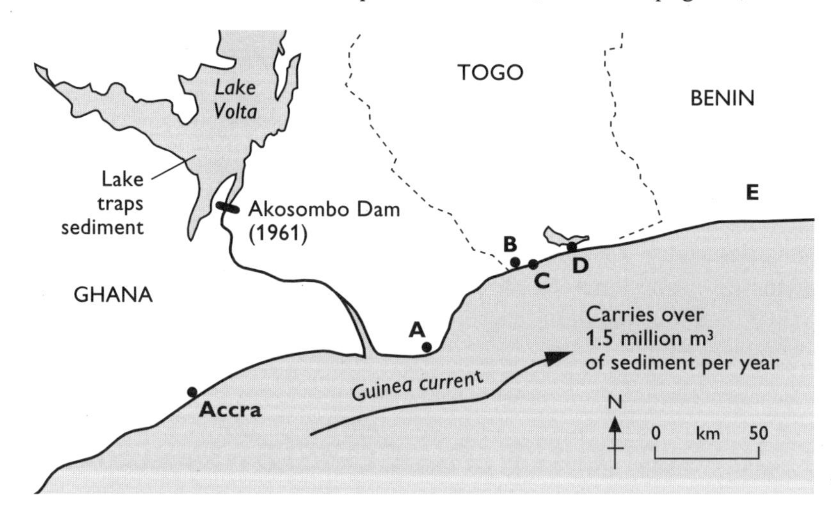


**Key**

**A** The town of **Keta** is falling into the sea because the beach is being eroded

**B** The breakwater at **Lomé** is preventing sand from travelling eastwards

**C** The holiday resort of **Tropicana** is losing its beach and its tourist income

**D** The jetty at **Kpeme** is being undermined (this port exports phosphates)

**E** The coastal oil wells of **Benin** are now being threatened

***Figure 31 The effects of human intervention in Ghana***

Similar interventions along the course of rivers may limit the natural development of river estuaries and deltas, or affect ecosystems like mangroves, sand dunes, salt marshes, mudflats and coral reefs, which also occupy coastal environments.

Our most common image of coasts remains one of beaches resting at the foot of cliffs. This model (see Figure 32) allows us to identify the different zones that make up the shoreline and, as we shall see in the next section, makes it easier to understand how physical processes change the features of beaches and cliffs.

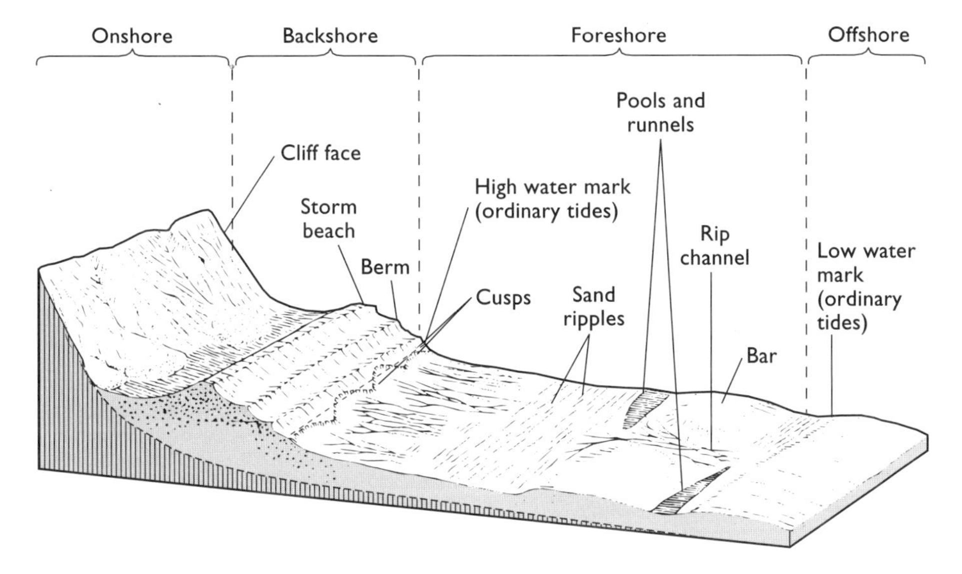


***Figure 32 Coastal zones and beach features***

# Short-term changes along the coast

Beaches occupy the **littoral zone** — the area between the highest and lowest spring tides. Here, waves arriving at right angles to the shore move sediment up and down the beach (**swash**), but if they arrive at an oblique angle, currents move material along the shore (**longshore drift**). These processes depend upon the shape of the coastline as well as the direction of the waves. They are regarded as short-term changes and show the dynamic nature of the coastal system.

Research suggests that sediment movements occur in distinct areas or **cells**, where inputs and outputs are balanced. Eleven of these have been identified around the coast of England and Wales. Sediment cells can be divided into smaller sections to allow closer study and management. Figure 33(a) shows the location of these cells and Figure 33(b) is a model of a smaller **sub-cell**, identifying its sources, transfers and sinks.

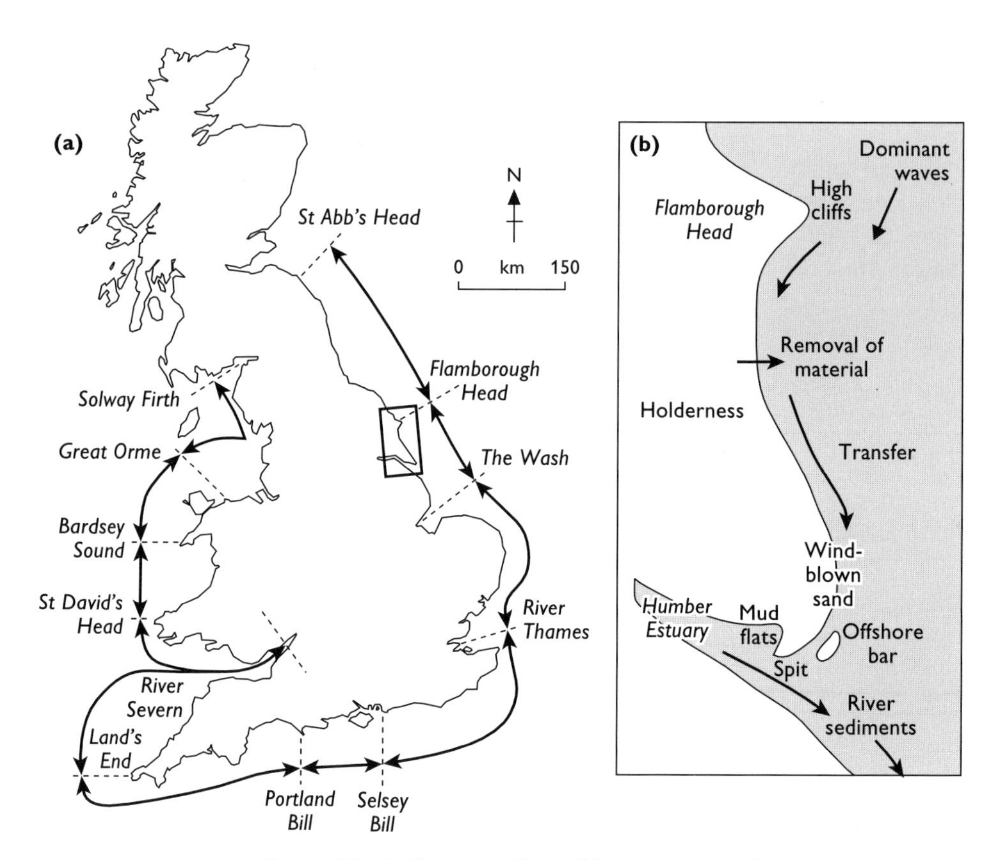


***Figure 33 Sediment cells and how they work***

Differing weather conditions and tides can also cause short-term changes, making waves either **destructive** (removing beach material) or **constructive** (building new features). **Storm activity** not only produces rapid coastal erosion and flooding, but also forms storm beaches and berms. Landslides and rock falls occur at such times in the backshore zone.

**Surge** conditions occur when storms and high tides coincide. Their effects may be further increased in enclosed and low-lying regions. In Bangladesh and India, the funnelling effect of the Bay of Bengal enables cyclones to flood large areas of the Ganges Delta (some 140 000 people died in these circumstances in 1991). Atlantic hurricanes off the coast of the USA can also drive seawater inland. In January 1953, storms caused increased erosion and flooding along the North Sea coasts of England and the Netherlands.

## Review 10

Using a copy of Figure 32 (page 41) and the ideas in this section so far, add arrows and annotations to draw up a model of a beach system, showing the likely inputs, transfers and outputs.

# Processes and change in coastal environments

## Factors influencing the nature of the coast

A wide range of factors, both physical and human, influences the nature of any coastline. Figure 34 divides these into four groups. This forms a kind of checklist of what shapes a particular coastline, and why it looks like it does. Many of these influences act together to produce characteristic features of erosion and deposition. You need to be able to identify coastal landforms and explain how they are being formed.

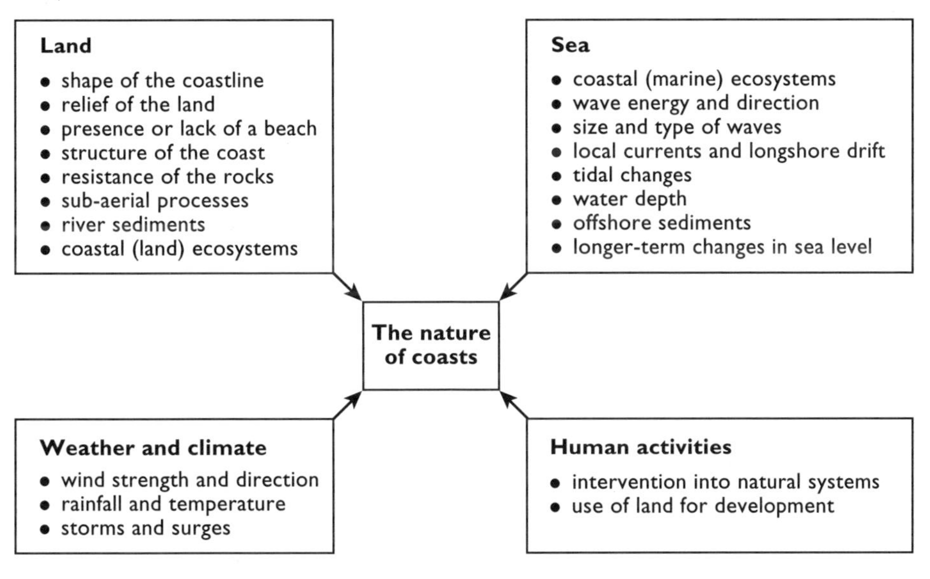


***Figure 34 The factors influencing coasts***

## The processes and landforms of erosion

All the factors shown in Figure 34 are important, but those relating to structure and lithology have a very strong influence on what a particular stretch of coastline looks like. There are a number of basic ideas that you need to know:

- the **lithology** or characteristics of rocks (especially resistance and permeability) varies — tough materials like granite and chalk tend to be eroded less by the sea, whereas weaker rocks like clay are eroded more rapidly

- variation in the rates at which rocks wear away is called **differential erosion**
- the **structure** or arrangement of the rocks (thickness and layers, etc.) affects the shape of a coastline — coasts with rocks lying parallel to the shore are described as **concordant** coastlines, and those with rocks at right angles to the shore are called **discordant** coastlines
- the **dip** or angle at which rocks lie can also increase or decrease the rate of erosion, as this affects how easily the waves can attack cliffs

## The Purbeck coast

In the western part of this area, the rocks run parallel to the coast and the resistant Portland limestone forms cliffs. These generally protect the coast from erosion and only in a few places has the sea broken through into the clay beyond. Lulworth Cove is perhaps the most impressive example of this. To the east the rocks are the same, but the situation is different. Here the structure is discordant and at Swanage the sea has been able to cut a bay in the weaker clays, creating limestone and chalk headlands either side.

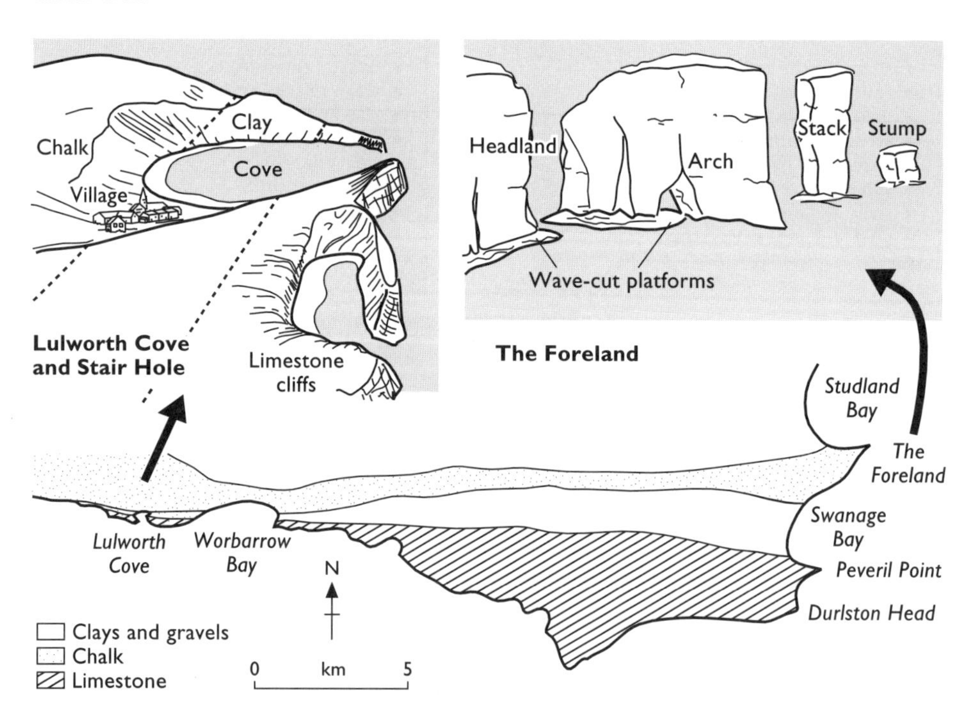


***Figure 35 The lithology and structure of the Purbeck coast***

## Review 11

Explain why the features of west and east Purbeck are different, despite having the same rocks.

The processes of erosion include the following:

- **abrasion** (corrasion) occurs when high-energy waves throw shingle at the base of cliffs. This rapid form of erosion undercuts the cliffs, forming a **notch** and causing rockfalls to occur. The cliffs thus wear back, leaving an eroded **wave-cut platform**.
- **hydraulic action** (wave quarrying) often occurs at high tide or when storm waves impact heavily on the **bedding planes** and **joints** in rocks. Air may become compressed in these weaknesses and so remove material rapidly.
- **corrosion** (solution) occurs when rocks like limestone are dissolved or become uncemented
- **attrition** occurs when sediments being moved by the sea gradually become more rounded and reduce in size because of erosion (beach materials have these characteristics)

In addition to these marine processes, there are also sub-aerial or land-based ones which help shape the coast. These involve **weathering** and **mass movement**. (There are links here to the Natural Hazards of Unit 5.) Mass movement includes the following:

- **rock falls** — for example, chalk cliffs collapse when they are being undercut by the sea
- **landslides** — material becomes wet and slips down along a lubricated layer
- **slumping** — saturated material moves suddenly, for example from clay cliffs onto the beach below
- **mudflows** — heavy rain can cause fine material literally to flow downhill

## The Holderness coast

Like Christchurch Bay or Lyme Bay on the south coast of England, the coast of Yorkshire shows how land and sea processes combine to shape the coastline. While this is seen famously in the cliffs at Robin Hood's Bay, Scarborough and Flamborough Head, it is in the Holderness section (shown in Figure 36 overleaf) that the most spectacular evidence is found. The soft clay cliffs made up of glacial material are retreating on average at over 1 m per year, and up to 15 times this rate in a few places. This coast has recorded the loss of 29 villages since Roman times.

On the cliff face, rainwash and **slumping** are key processes, while at the cliff foot the narrow beach is unable to prevent waves removing the clay (in suspension). Storm and surge conditions often cause spectacular damage. Longshore drift moves on average almost 1 tonne of sediment per minute southwards.

The most important depositional feature here is the 6 km-long **spit** at Spurn Head. Although it is growing at about 10 cm per year, winter storms do periodically threaten to detach it from the mainland. The Humber Estuary interrupts the coastline here and it is in this more sheltered environment that sand dunes, saltmarshes and other ecosystems are found.

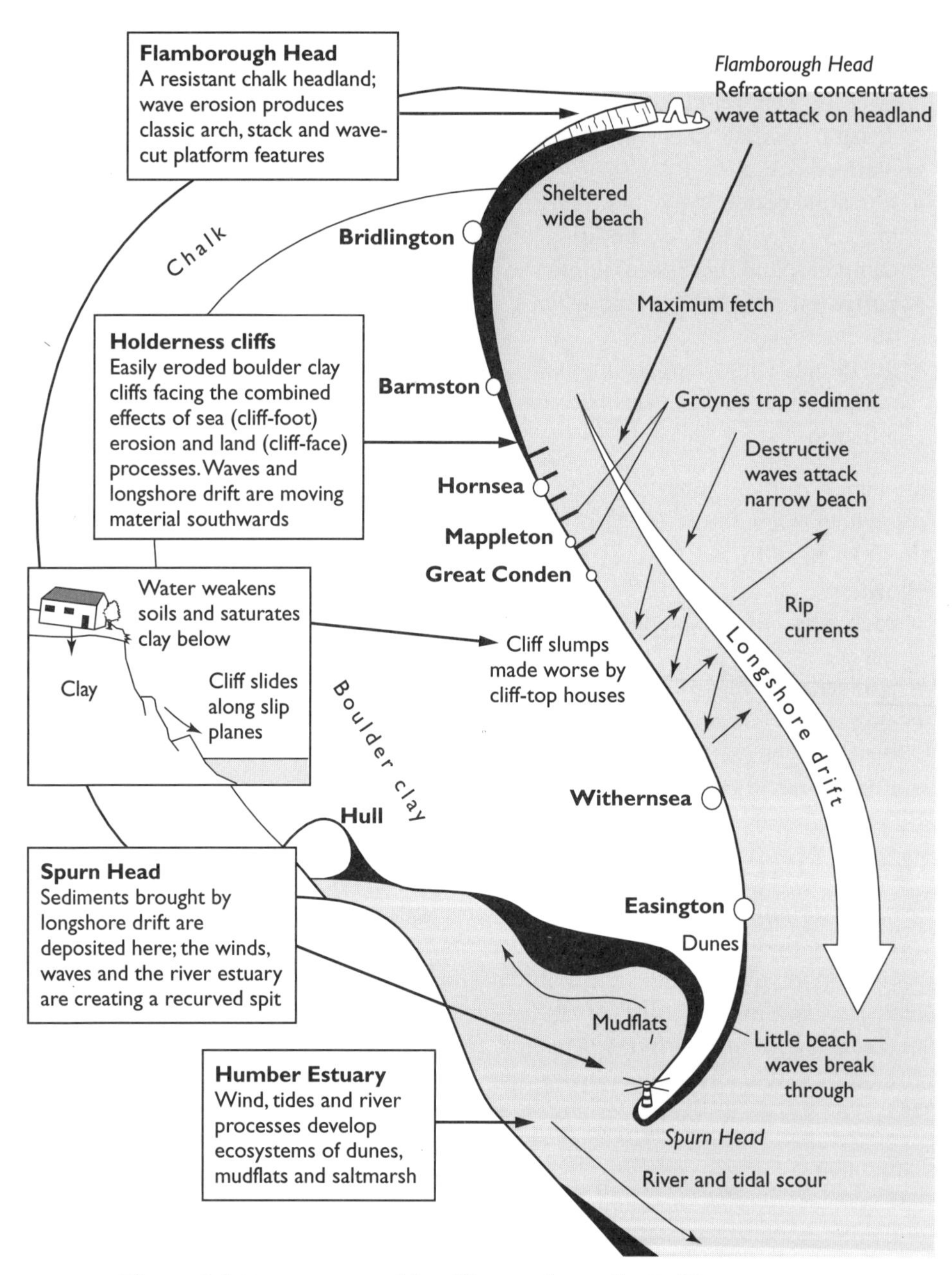


***Figure 36  Processes and landforms along the Holderness coast***

## Review 12

Using two annotated cross-sections, compare the cliff profiles and processes at work on a chalk headland (Flamborough or The Foreland) and a clay bay (Lulworth or Holderness).

# The processes and landforms of deposition

Theory says that there are two common beach forms — swash-aligned and drift-aligned.

**Swash-aligned** beaches occur when waves arrive at right angles to the shore, or are turned shorewards by frictional drag. This process is called **wave refraction**. These beaches have distinctive parallel features that reflect the processes at work (they can be seen in Figure 32 on page 41):

- offshore bars of sand, combed from the beach by the waves, form near to low-tide level
- detailed features like runnels, rip channels and beach cusps result from localised water movements on the beach
- where the sediment size is larger, the beach profile becomes steeper and percolation is faster
- large berm ridges are formed by waves at high-tide level
- winter storm waves can create steep beach features backshore

**Drift-aligned** beaches occur when waves arrive at a narrower angle to the shore. This angled approach and steep wave return produces a sideways movement of water and sediment that we call **longshore drift**. **Spits** like Spurn Head result where this process dominates. In particular circumstances, these landforms can develop further into other features:

- waves arriving at 30° create the greatest amount of drift
- spits occur when the coastline is interrupted by a change in the direction of the coastline or by a river mouth (e.g. at Orford Ness in Suffolk)
- wave refraction, tidal currents or changing winds can **recurve** spits shorewards (e.g. Hurst Castle)
- **tombolos** are spits that link islands to the main coastline (e.g. Chesil)
- bay **bars** result when spits grow across a bay and become attached at both ends (e.g. Looe bar in Cornwall)
- **barrier islands** are larger features which have developed along coastlines (e.g. along much of the Atlantic and Gulf coasts of the USA)

In practice, there remains some dispute about how the larger features may have formed, and modern theories suggest more complex origins.

The US barrier islands shown in Figure 37 overleaf are seemingly formed by storm waves periodically washing water and sand over the islands. This causes the islands to migrate landwards. Their formation may also be related to the effects of rising sea level or to the moving of offshore sediments (left there after the Ice Age, and now being recycled by present-day waves). This migration process creates problems for the people and businesses located there, as illustrated in Figure 41 (page 54).

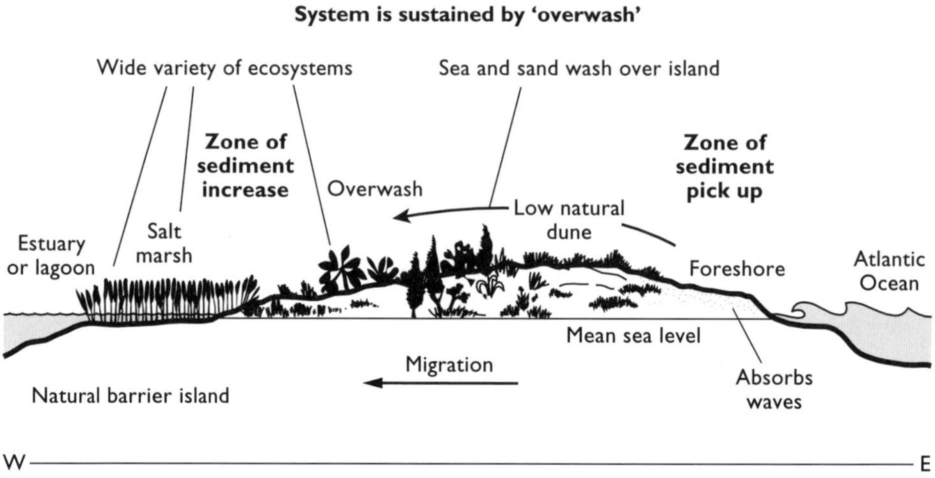

***Figure 37 The formation of barrier islands off the eastern USA***

# Coastal changes and time

Changes along coastlines may be short term or longer term, ranging from brief, localised storm events, to large, global changes occurring gradually, perhaps over geological periods of time. The six examples shown in Table 3 illustrate this range, and their impacts are looked at in the next section.

***Table 3 Examples of coastal changes***

| Location | Timescale | Cause | Changes |
|---|---|---|---|
| Towyn, North Wales | Days following 26 February 1990 | Intense low-pressure storm, onshore winds, tidal surge | Sea breached sea wall and flooded coastal homes |
| Holderness, Yorkshire | Since Roman times | Erosion of clay cliffs by land and sea processes | 5 km retreat of coastline destroying 29 villages |
| Barrier Islands, eastern USA | Since last Ice Age | Repeated overwash of storm waves and rising sea levels | Barrier Islands are migrating landwards |
| Scotland — all land once covered by ice | Since last Ice Age | Isostatic — land recovers regionally now that weight of ice has gone | Land rising relative to the sea |
| Mauritius island and nearby coasts | Since last Ice Age | Eustatic — global warming means that ice is melting and sea expanding | Sea level rising globally, affecting lowland areas |
| California, USA | Over 15 million years | Tectonic — mountain uplift near to plate boundaries | Land rising regionally |

## Long-term changes

**Positive** movements of sea level are when water rises relative to the land, having a drowning effect upon the coast. This has happened since the last Ice Age, and it is predicted that the melting ice caps will lift sea level by a further 50 cm this century. **Estuaries**, **rias** and **fjords** are created in this way, making excellent harbours and offering opportunities for development. Rising sea levels can, however, have costly and hazardous impacts along lowland coastlines, especially if they are heavily populated or developed.

**Negative** changes in sea level allow new coastlines to emerge from the sea. While this can be caused by falls in sea level — about 140 m during the last Ice Age — the more contemporary cause of this change is **isostatic recovery**. Since the last Ice Age, many parts of the world, freed from the enormous weight of ice, have begun to 'rebound' and so rise relative to sea level (currently up to 4 mm per year in Scotland). Fossil cliffs and **raised beaches** are clear evidence of this coastal uplift, and the land gained, though often small in area, is of great value to some coastal communities.

# Coastal ecosystems

Coastal ecosystems reflect the nature of the environments in which they occur. Within the tropics, the mangroves and coral reefs are important, while in the UK sand dunes (**psammoseres**) and salt marshes (**haloseres**) dominate. Being at the boundary between land and sea, these ecosystems must adapt to the changing conditions (e.g. water depth, salt content and wind speed). Unfortunately, such locations are also increasingly in demand for recreational, agricultural and industrial development, and are therefore at risk.

## Plant succession

**Plant succession** is the process by which vegetation develops as it colonises areas and becomes increasingly adapted to the changing environment. This process has characteristic features:

- Bare ground or shallow water gradually becomes **colonised** by plants. **Pioneer species** are able to invade because they can cope with the tough conditions of these abiotic environments.
- As the environment is modified, so new plants become established. This continuing change is called the **prisere** or primary sere.
- Each new community of plants forms a separate **seral stage**.
- Eventually a balance is reached between plants and the environment, and the resulting plant species form a **climatic climax community**.
- Often, however, this succession is interrupted by **arresting factors** that stop development.

- If the factor is natural (e.g. salinity or relief), the result is a **sub-climax community**.
- If the control is human in origin (e g. agriculture), it is referred to as a **plagioclimax**.
- If an arresting factor is removed, the succession process may restart, but if this happens a more limited succession occurs.

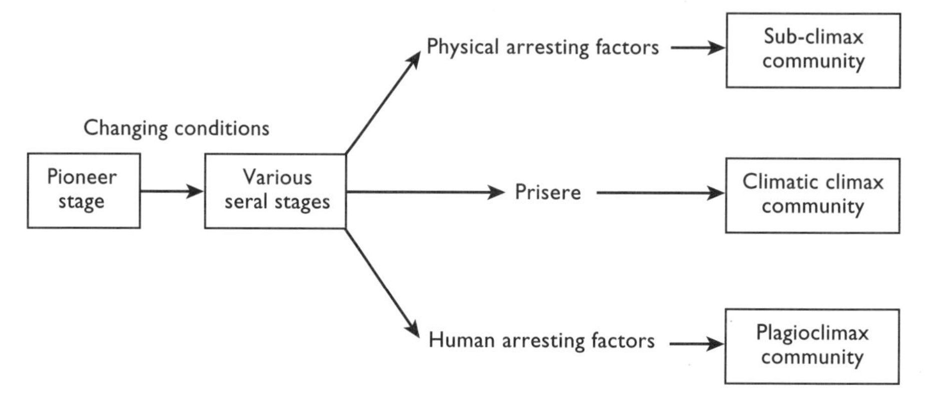


***Figure 38 Ecological plant succession***

## Sand dune ecosystem

The formation of a sand dune succession (**psammosere**) requires certain environmental conditions:

- a fresh supply of sand, usually from an adjacent wide, sandy beach source
- a frequent onshore wind direction
- vegetation, often in the form of marram grass, to give some stability to the growing dunes
- protection from storms and damaging human activity

Ainsdale on the coast of Merseyside is famous for its sandy beach, sand dunes and rare natterjack toads. The dunes here cover some 800 hectares, most of which is a nature reserve. There is, however, increasing human pressure from recreational activities, such as increased pedestrian and cycle traffic from the nearby holiday resort, and from pine and birch tree invasion, planted here in the past to help stabilise the dunes. There has also been some recent coastal erosion, and inland there is housing, golf course and road development. All of these changes are preventing the natural succession and dune migration.

Figure 39 summarises data that have been collected along a transect through the Ainsdale dunes. The method used samples of vegetation (diversity and cover) taken by means of quadrats at points along a line perpendicular to the shore. These surveys are usually repeated along the dunes to get a fuller picture of what is happening. Other variables, such as dune height, soil characteristics and wind speed, are also measured. You should carry out or at least know how to carry out this type of investigation (see *Geography Review* Vol. 14, No. 4, pp.14–17). (See also **www.merseyworld.com**)

| | Foredunes | Mobile dune (yellow) | Fixed dune (grey) | Dune slack (fresh water pool) | Dune heath |
|---|---|---|---|---|---|
| Sea ← | | | | | |
| % bare ground | 100 | 85 | 10 | na | 0 |
| No. of different species | 1 | 5 | 15 | 27 | 12 |
| Most dominant (%)<br>Other common plants | Sea couch grass (1) | Marram grass (90)<br>Red fescue | Sand sedge (75)<br>Marram, other grasses, creeping, willow and dewberry | Rushes, reeds, grasses, mosses, willow, irises, rare natterjack toads and orchids | Ling heather (90)<br>Heath rush, gorse, broom, buckthorn, (some pine and birch) |
| Soil | Sand and shells, salty, very dry, pH 8 | Sand with humus, still dry, pH 7 | Brown humus layer, some water, pH 5 | Winter saturation | Black deep humus, holds water, pH 4 |
| Recent human changes | Supply of blown sand decreasing, embryo dunes not forming | Trampling, path erosion and gullies; dunes becoming 'fixed' | Damage along trails and foot-paths; 'blowouts' can occur if vegetation is removed | Scrub is invading the slacks, drying them out in summer | Plantation (pre-war) growth and access by public need to be controlled |

***Figure 39 A transect diagram across Ainsdale sand dunes, Merseyside***

## Review 13

(1) Redraw Figure 38 as a psammosere, adding details taken from Figure 39.

(2) Explain how and why conditions, and therefore plants, change across the Ainsdale dunes.

(3) Why, despite the pine and birch trees, has this become a plagio- rather than a climatic climax?

(4) Suggest how you might try to manage this area better.

***Tip*** Think about land management (repairs, upper beach nourishment, land use and planning controls, etc.), habitat management (planting, herbivore controls or even introduction, scrub removal, etc.) and visitor management (signposts, reserves, education, etc.).

## Coral reefs

Along tropical coastlines, coral reefs support a rich marine diversity, fish production and valuable tourism, as well as being a natural protection from the sea. Some experts say they may yet turn out to be ecologically as valuable as tropical rainforest.

The traditional lifestyles of subsistence farming and fishing in island communities, such as the Maldives, only had a limited effect on coral reefs but, increasingly, reef **atolls** are being threatened by human activity. These threats are both local (from development) and global (from rising sea levels). Remember that coral will only grow in conditions of warm, light, clear (silt free) and relatively shallow salt water.

**Coral reef exploitation** needs to be better managed. The problems include:

- overfishing of reefs and lagoons
- removal of beach material and lime (coral) for building
- dredging (even dynamiting) channels to improve harbour access
- reclamation of lagoons and the cutting of the mangroves behind
- use of the reefs by holiday visitors for recreation (diving)

**Rapid population growth** among many tropical island communities is increasing the need to develop the limited land available, leading to:

- clearance of inland forest cover for agriculture, which will increase freshwater and sediment input
- greater use of agricultural chemicals, which may pollute rivers
- pollution from larger-scale industrial processing plants
- increased quantities of commercial and domestic sewage, causing eutrophication
- removal of mangroves to increase fish farm production

**Natural and enhanced global warming** means:

- a faster rate of rise in sea level
- an increasing likelihood of tropical storms
- higher water temperatures, which cause bleaching and death (an El Niño effect)
- increased carbon dioxide, which makes shallow water more acidic

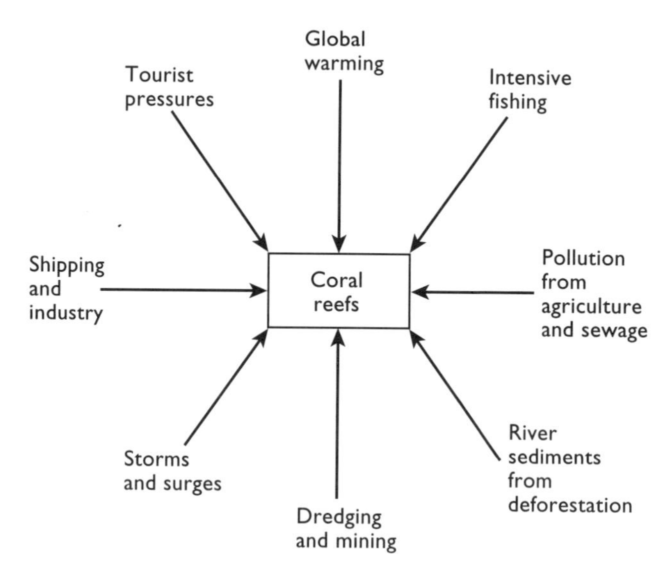


***Figure 40 Threats to coral reefs***

***Tip*** You will have studied and named coral reefs, so you need to be able to state examples for each of these threats, and to classify the threats as short- or long-term and as physical or human.

# Environment–people interactions in coastal environments

## The impact of short-term changes

The effects of change can be seen in a variety of coastal environments. **Coastal recession** and **flooding** create the most issues and conflicts. Some relate to physical factors, whereas others are caused by human activity. Here are some examples that you could research further.

### The Holderness coast of Yorkshire: the impacts of rapid erosion

- Evidence of 29 villages being destroyed since Roman times.
- Mappleton church was once 3 km from the sea; now it is less than 300 m.
- The cliffs south of Withernsea have retreated 212 m in the last 100 years.
- Spurn Head is migrating westwards and winter storms threaten to breach the neck.
- Flood defences along the Humber floodplain and estuary are under pressure.
- Channel migration and sediment deposition are recurring problems in the estuary.

### Towyn, North Wales: short-term coastal flooding

- On 26 February 1990 a deep depression (951 mb) drove 70 knot winds onshore.
- At high tide, storm waves first overtopped and then later breached the sea wall.
- This event was calculated to have a 1 in 500 **recurrence level**.
- Was the sea wall, also a railway embankment, strong enough?
- Should housing, especially bungalows, have been built behind the wall?
- Five thousand, mostly elderly, people were evacuated with difficulty.
- Caravans and other holiday resources were badly damaged.
- Many people had no home contents insurance.

### The US barrier islands: inland migration of sand

- The barrier islands stretch 2700 km from Maine to Texas.
- They are a valuable defence against periodic hurricanes and coastal erosion.
- They are migrating landwards (4 m per year at Cape Halteras, South Carolina).
- Rising sea levels are increasing the problem.
- In some locations, such as Florida, investment in development has been massive.
- Preventing the retreat of the islands by hard engineering is not a long-term solution.

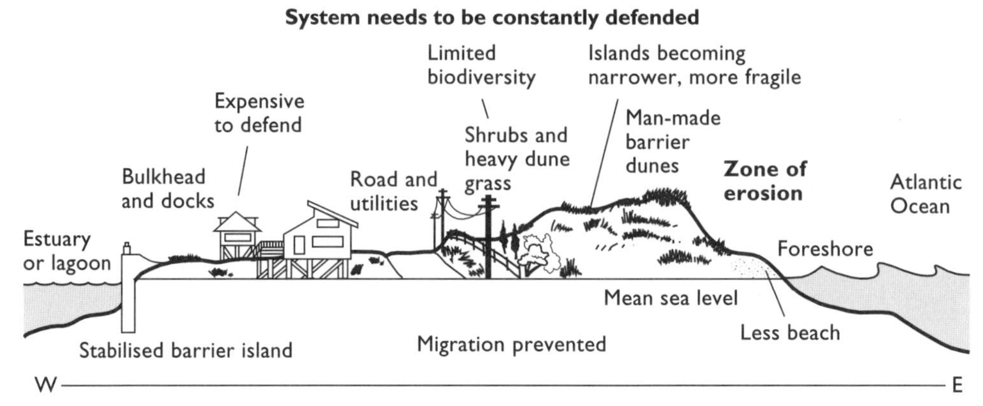


***Figure 41 A developed section of the US barrier islands (compare this with Figure 37 on page 48)***

# The global impact of human activities

Throughout the world, coasts are coming under increasing pressure from human activities. The land in coastal regions is 'overcrowded, overdeveloped and over-exploited' (Don Hinrichson, *Living on the Edge*, 1995). The global extent of this pressure is shown in Figure 42.

**Overcrowded**

- By 2025, 75% of the world's people (6 billion) will live in coastal areas.
- In China, 100 million people have already migrated to the coastal cities.
- Population densities in Florida and southern California will double by 2025.

**Overdeveloped**

- The Atlantic coasts of mainland Europe and the USA are almost totally urbanised.
- The Mediterranean has as many summer tourists as it has residents.
- Bombay puts 1 million tonnes of sewage and waste into the Indian Ocean each day.

**Over-exploited**

- Most of the world's inshore fishing industries have collapsed.
- Half of the world's mangrove forests have been felled for wood or fish farming.
- Only 70% of the world's coral reefs remain stable.

## Review 14

(1) Use an atlas and the information from this section to annotate Figure 42, to show some key coastal issues/problems, and the locations involved. Some issues related to sea level rise are shown already.

(2) To what extent do you think that these problems are human rather than natural in origin?

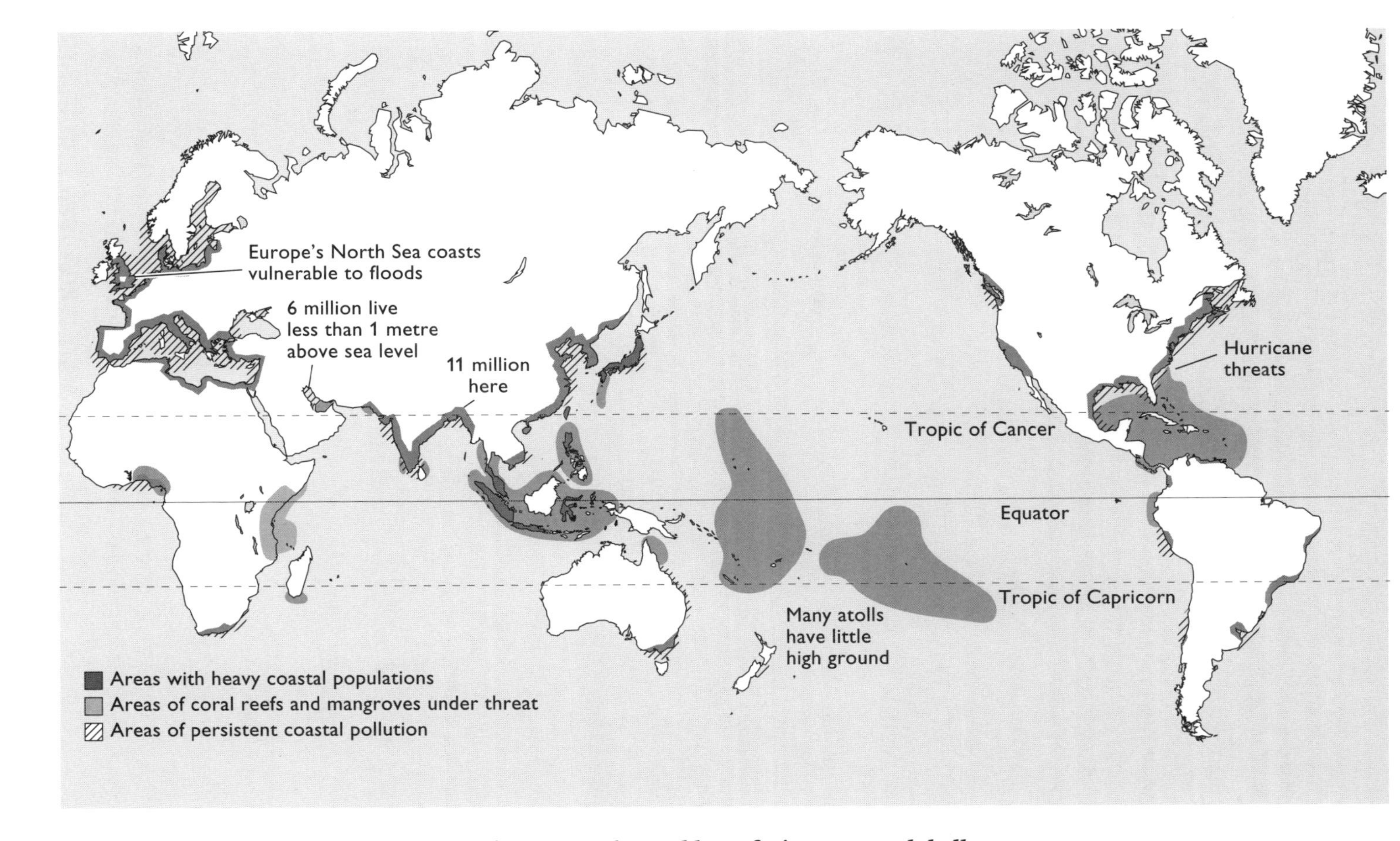


*Figure 42 The problems facing coasts globally*

## South and east coast USA

The consequences of human activities along any coastline are often related to:

- urban and industrial development and water quality
- recreational and tourism pressures
- land reclamation (agricultural) schemes

They occur in both MEDCs and LEDCs, although the eastern coastline of the USA reveals many of them.

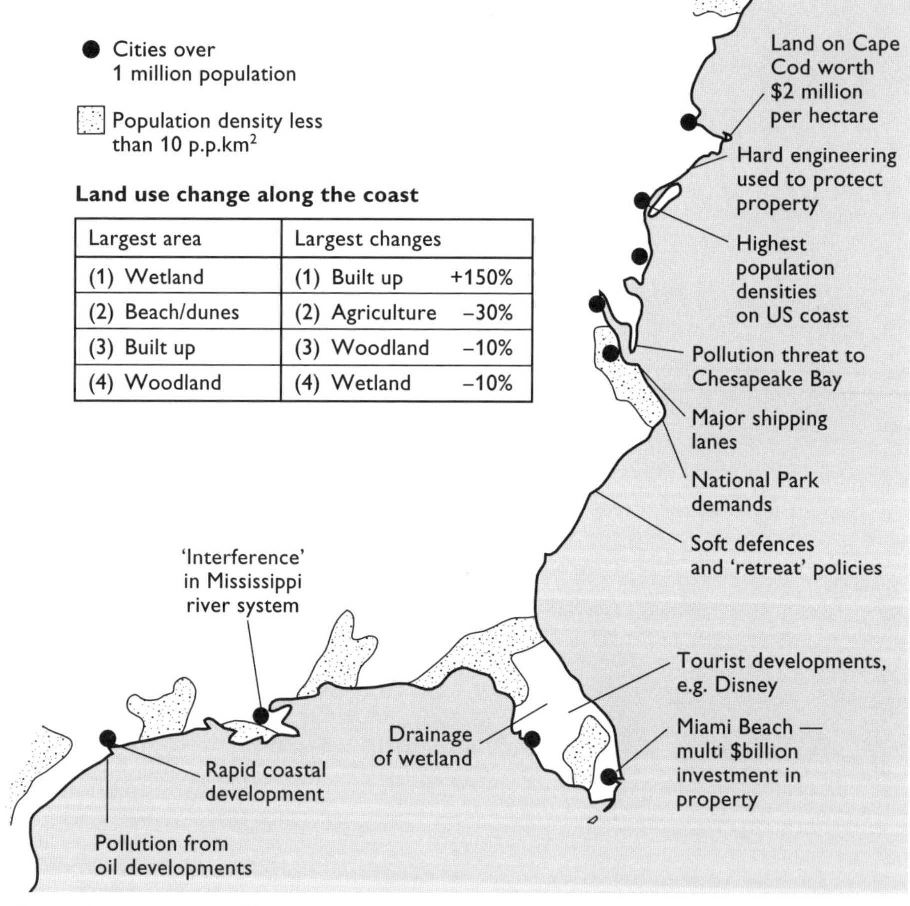


| Largest area | Largest changes |
|---|---|
| (1) Wetland | (1) Built up +150% |
| (2) Beach/dunes | (2) Agriculture –30% |
| (3) Built up | (3) Woodland –10% |
| (4) Woodland | (4) Wetland –10% |

***Figure 43 Impacts of human activity along the south and east coast of the USA***

These developments are valuable economically, but they can create problems and issues that might lead to conflict between the developers and the protectionists:

- the developers in the USA include government (planning, ports, state and federal), heavy industry (oil, manufacturing and utilities), business (builders, real estate and insurance) and recreation (hotels and leisure)
- the protectionists might include environmental groups (Sierra Club, Worldwide Fund for Nature), residents' groups and fishing interests

A **conflict matrix** is a grid that allows you to examine how various people or groups may hold compatible or conflicting views or objectives. Boxes can be ticked as compatible or crossed as conflicting. They may even be awarded scores.

***Table 4 A conflict matrix***

| | Fishing interests | Ports agency | Oil company |
|---|---|---|---|
| Fishing interests | | | |
| Ports agency | ✓ | | |
| Oil company | ✗ | ✓ | |

## Review 15

Explain the conflicts shown in Figure 44.

***Figure 44 The battle for Britain's beaches***

# Coastal management and the future

## Approaches to coastal management

There is clearly need for careful management to protect coastal areas and the people and resources there. Management strategies cover a wide range, from major projects involving massive civil engineering, to small-scale work supporting coastal ecosystems. The decision to 'do nothing' may even be the most appropriate strategy. The strategies are summarised in Table 5.

***Table 5 Management strategies — hard and soft options***

| **Strategy** | **Purpose/description** | **Strengths** | **Weaknesses** |
|---|---|---|---|
| HARD ENGINEERING | OVERCOMING NATURAL PROCESSES | | |
| *(1) Cliff-foot strategies* | *To protect the cliff foot or beach from sea erosion* | | |
| Sea walls | Massive, made of concrete, increasingly used to reflect rather than prevent waves | Traditional solution to protect valuable or high-risk property, dense population or resources | Very costly, foundations can be undermined if built on beaches, or where longshore drift operates |
| Revetments | Massive, made of rocks or concrete, absorb waves, are porous or act as baffles | As sea walls, but relatively cheaper | Do not cope well with strong storm waves |
| Gabions | Smaller rocks held in wire cages | Have some properties of both of above, yet much cheaper | Relatively lightweight and small-scale solution |
| Groynes | Hold beach material threatened by longshore drift | Low capital cost and repaired easily | Cause scour down drift, have wider impacts on sediment cells |
| Offshore breakwaters | Reduce power of waves offshore | Can be built from waste material, mimic reefs | Possible ecological impacts and may not work on large scale |
| Rip-rap (rock armour) | Large rocks at foot of sea walls or cliffs to absorb waves | Effective, cheaper than revetments, prevent undermining | May move in very heavy storm conditions |

| Strategy | Purpose/description | Strengths | Weaknesses |
|---|---|---|---|
| *(2) Cliff-face strategies* | *To reduce damage to cliff face from sub-aerial erosion* | | |
| Cliff drainage | Removal of water from rock prevents landslide or slumping | Cost-effective | Drains can be new weaknesses, dry cliffs can cause rock falls |
| Cliff regrading | Lower angle of cliffs to prevent collapse | Works on clay material, when little else will | Retreat of cliff line uses up large area of land |
| SOFT ENGINEERING | WORKING WITH NATURAL PROCESSES | | |
| Offshore reefs | Mining waste or old tyres fastened together and sunk, acting like wave 'speed bumps' | Relatively cost-effective and low technology | Largely untested beyond research level and may have pollution implications |
| Beach nourishment | Sand pumped from seabed to replace eroded beach | 'Natural'-looking process | Very expensive, may erode again or have ecological impacts |
| Managed retreat | New buildings and defences are prevented, incentives given through grants etc. | Cost-effective, preserves natural coastline and probably saves lives | Difficult to persuade people they are safe, or that the authorities do care about them |
| 'Do nothing' | Accept that there is no economically viable or technically feasible solution | Cost-effective, allows chance to research or await new techniques | Unpopular locally, political implications |
| Red-lining | Planning permission withdrawn and new line of defence set up inland — 'set-back' schemes | Cost-effective | Unpopular locally, political implications |

There is some argument about terms like 'soft engineering', and it should not be assumed that such techniques are necessarily sustainable or even cost-effective. What is important is to decide whether a strategy is appropriate for the problem and location concerned. You need to produce a simpler version of Table 5 that relates to the coast you intend to study.

# Decision-making about coasts

You should understand four important ideas:

- **Cost–benefit analysis (CBA)** is used to decide if a project is **economically viable**. It compares the costs of building and maintaining a scheme against the benefits gained (property and infrastructure saved, facilities, jobs and business gained). In practice, the benefits total is divided by the costs, and a ratio of over 1 may suggest that a scheme is worth doing. It is difficult to put a price on amenities and people's personal values, but CBAs are becoming increasingly sophisticated, trying to allow for lost opportunities, inflation and even some environmental effects (e.g. Miami Beach nourishment: cost $67 million, resort assets $5 billion).

- **Feasibility studies** look at geology and coastal processes from an engineering point of view to help decide whether a particular scheme is appropriate for that location. Such research is very important.
- **Environmental impact assessment** is carried out when there is a risk that natural ecosystems or local environments will be affected by developments. Some attempt is made to assess the impacts. The presence of sites of special scientific interest (SSSIs) or fragile ecosystems may lead to schemes being modified or even rejected.
- **Risk assessment** involves looking at the likely *recurrence of events* and the value of what is at risk (e.g. lives and property). Insurers and lawyers are increasingly using this type of calculation. Changes expected as a result of global warming will make this approach to planning even more likely in future. The Environment Agency frequently adopts this approach with flood warnings.

# Contrasting coastal management policies

The way in which coastal management is operated varies from country to country. In LEDCs such as Bangladesh, the strategies chosen must take into account the country's limited finances, and so tend to be about dealing with consequences. The risks to life are great and the choices limited. In MEDCs, attempts are made to prevent problems. There is access to more sophisticated technology, but the costs of damage in financial terms are much greater.

## Coastal management in the UK

Historically, management in the UK has been a very piecemeal affair, often in the hands of local councils, non-government organisations and even private individuals. This has led to conflict and situations where one authority has built defences that cause damage somewhere else. Typical of this is the building of groynes, which create beaches in one resort yet effectively destroy them in another. Barton on Sea is a classic case of this problem.

The current policy is towards more **integrated management**. Since 1995, the Ministry of Agriculture, Fisheries and Food has encouraged groups to work together to produce **Shoreline Management Plans** or SMPs (see Figure 45), based on sediment sub-cells (see Figure 33(b) on page 42). These SMPs are voluntary at present, but could well be how all future decisions are taken. The increasing role of the Environment Agency may help to police coastal activities and bring conflicting groups together.

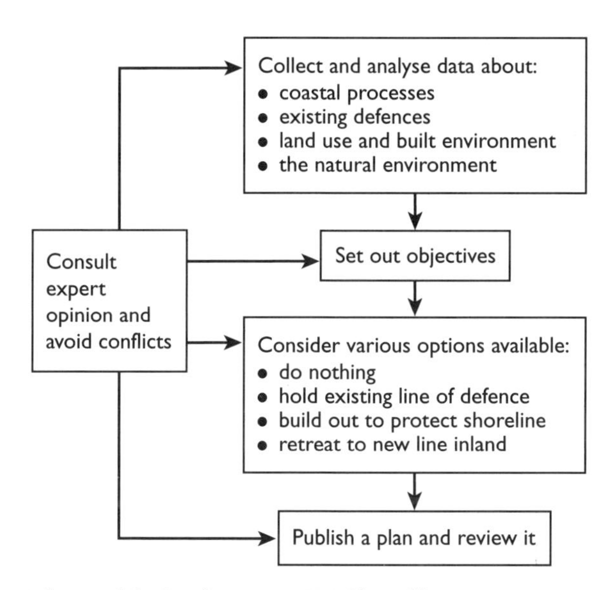


***Figure 45 Setting up a UK Shoreline Management Plan***

## Coastal management in the USA

Since 1972, the policy in the USA has been to set up **Coastal Zone Management Programs**. These are run by each state through its Division of Coastal Management (www.nos.noaa.gov/ocrm/czm), with support from federal funds via the Water Resources Development Act. They have three responsibilities/strategies:

- to attempt to **control** erosion or flooding by using hard or soft engineering. Virginia Beach, for example, has massive sea walls that protect this ageing seaside resort, but which have also caused erosion of the beach. Monmouth Beach has one of the nation's largest beach nourishment schemes. Rising sea levels and demands from other sectors of the economy mean that it is not easy to justify spending huge sums of taxpayers' money on coastal defences.
- to **accommodate** problems using the National Flood Insurance Program. This offers insurance only to those who build away from areas of high flood risk and refuses new defences for those who do not. Many still take the risk, however, as is seen in the growth of places like Ocean City, Maryland.
- to **adjust** to problems by land use zoning or set-back laws. These controls are unpopular, as they lead to conflict. They prevent development in high-risk areas, but will compensate existing users who are prepared to make a 'strategic retreat'. The Massachusetts Wetlands Act, for example, not only limits new developments on the coast, but actually bans any form of permanent coastal defence.

It seems that the UK, along with many other MEDCs, is moving towards this American style of coastal management, which includes integrated programmes, public acquisition of wetlands and more use of private funds. An exception to this trend is the Netherlands, where there is direct government funding.

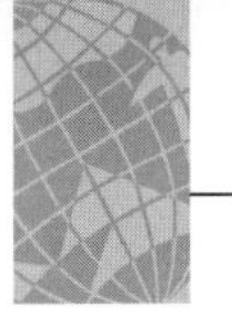



# Sustainable coastal management — a way forward?

Decisions about where to implement defences are becoming tougher, and such defences are often recommended only when 'in the national interest' (e.g. to protect power stations or major urban areas). Elsewhere, more 'managed retreat' will mean that farmland will return to saltmarsh, improve wildlife habitats and provide natural protection. Small-scale **sustainable management** thus seems possible, even desirable.

In the Netherlands, there is currently a major commitment to hard engineering, as much of the country is below sea level. However, there is a growing realisation that this strategy is untenable in the long term, and that more sustainable schemes are also required. New Dutch ideas about 'growing with the sea' are exploring the possibility of **coastal resilience**. This suggests that if parts of the coast were allowed to be flooded by the sea, ecosystems like saltmarshes and wetlands would be able to adapt to rising sea levels. Similarly, if selective coastal erosion were allowed, sand dune development might provide protection. Whether schemes on this scale can succeed, given their high risks, is less certain.

## Review 16

(1) Explain the differences between:
   (a) cliff-foot and cliff-face strategies
   (b) hard and soft engineering
   (c) cost–benefit analysis and environmental impact assessment

(2) Research one example of sustainable coastal management.

# Questions & Answers

In this section of the guide there are three questions based on the topic areas outlined in the Content Guidance section. Each question is worth 30 marks. You should allow 30 minutes when attempting to answer a question, dividing that time according to the mark allocation for each part.

The section is structured as follows:

- Three specimen questions are presented, similar in style to those in the Edexcel B papers. These illustrate typical questions (2), (3) and (4) (see page 5). Remember, there will be five to choose from in the real examination.
- Answers have been provided to the short-answer questions, showing you how to squeeze the maximum marks out of data response in order to achieve an A grade.
- For the long-answer questions, an answer of A-grade standard and an answer of C-grade standard have been provided so that you can see the reasons for the difference in achievement. Using examples and case studies is a vital part of these long-answer questions, and the A-grade standard answers show you how to do this.

**Examiner's comments**

Candidate responses to long-answer questions are followed by examiner's comments, preceded by the icon *e*. They are interspersed in the answers and indicate where credit is due. In the weaker answers, they also point out areas for improvement, specific problems and common errors such as poor time management, lack of clarity, weak or non-existent development, irrelevance, misinterpretation of the question and mistaken meanings of terms.

The comments indicate how each example answer would have been marked in an actual exam. For questions with only a few available marks, examiners generally give each valid point a mark (up to the maximum possible). For higher-mark questions, examiners use a system involving 'Levels':

*Level I* At this level, answers generally contain simple material. Points are stated briefly with no development. Examples are given in one or two words, usually in the form 'e.g....', and explanation is basic, such as 'A is the cause of B'.

*Level II* Answers at this level need to contain more detail. For instance, if the question requires a comparison, a Level II answer will contain some quantification, such as 'A's GDP is 10 times larger than B's'. Examples are more detailed and explanations involve clearer reasoning: 'A is the cause of B because...'.

*Level III* At AS there are only a few occasions when an answer is marked at Level III, usually when 9 or 10 marks are available. Marks are awarded for breadth and depth, for drawing together the threads of a response and for overall evaluation.

# River environments

**(a) Study Figure 1. It shows Schumm's model of how river channel variables can change from source to mouth.**

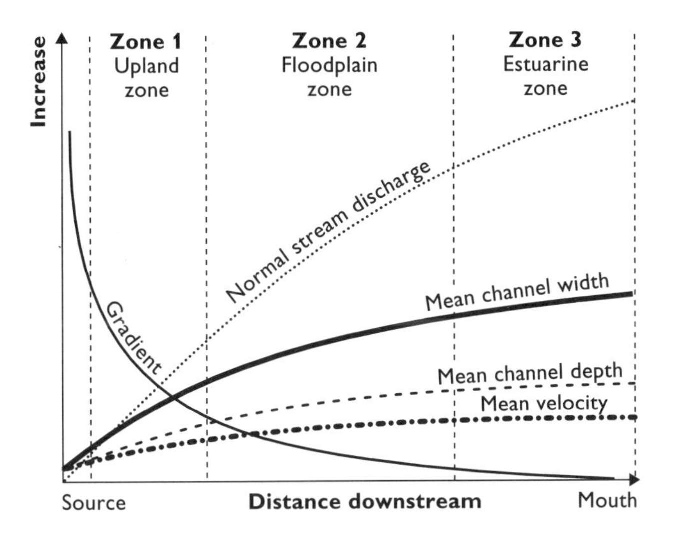


*Figure 1 Changes in channel characteristics from source to mouth*

**(i) Define the following terms:**
- **velocity**
- **discharge** (3 marks)

**(ii) Describe the changes shown in velocity and discharge from source to mouth.** (3 marks)

**(iii) Give reasons for the changes shown in velocity and discharge from source to mouth.** (7 marks)

**(b) Study Figure 2. It shows how non-regulated building can spread on floodplains.**

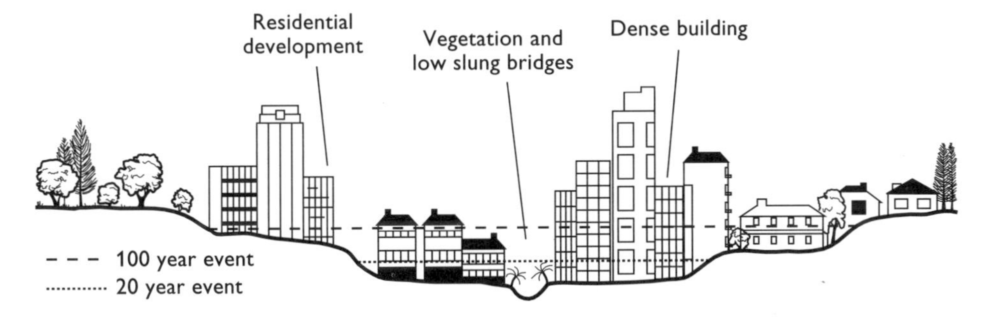


*Figure 2 Spread of non-regulated building on floodplains*

**(i) Suggest how the floodplain developments shown can increase the risk from flooding.** (3 marks)

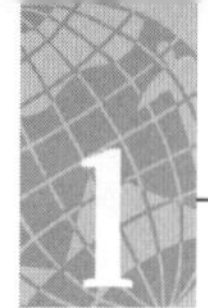




**(ii) Explain two ways in which you could modify the river channel to cut down the risk of flooding.** (4 marks)

**(c) Recently, more sustainable management strategies are being used to manage rivers. Describe some examples of sustainable strategies and examine their advantages and disadvantages.** (10 marks)

■ ■ ■

## Answers to parts (a) and (b)

**(a) (i)**
- Velocity is the speed of the river flow: distance travelled per unit of time.
- Discharge is the amount of water passing a particular point (through a cross-section) over a unit of time (calculated as velocity × volume).

*e* 1 mark for (a), 2 for (b).

**(ii)**
- The mean velocity gradually increases from source to mouth, but at a diminishing rate nearer the mouth.
- The discharge increases dramatically from source to mouth.

*e* 2 × 1 mark for the basic trends, with a bonus mark for any finer point.

**(iii)** Velocity
- Although the gradient decreases, the decrease in channel friction more than compensates.
- There is increased efficiency (hydraulic radius).
- There is a larger volume of water in relation to wetted perimeter (drag from banks and bed).
- There is a smoother channel in the lower course, so less friction or less turbulence.

Discharge
- The velocity increases (see (ii)).
- The volume increases because width × depth increases.
- Tributary streams join together to increase discharge.

*e* One mark is available for each basic reason (e.g. the first discharge answer), while 2 marks are available for a fuller explanation (e.g. the last two velocity answers). The overall mark for (iii) is 7, with these marks being scored across both parts of the question (e.g. 4+3 or 5+2).

**(b) (i)**
- The incidence of high-value buildings and dense settlements increases risk of damage.
- The bridges can be choked by debris and water will back up.
- Increased urbanisation leads to more impermeable surfaces, less infiltration and more surface flow.
- Urban drains and gutters produce faster and higher flood discharge peaks.

*e* 1 mark for each, up to a maximum of 3.

(ii)
- Dredging — wider and deeper channel.
- Re-sectioning — wider channel or smoother channel.
- Realignment — straightening of course encourages faster dispersal of water.
- Building higher banks (on top of natural levées) can hold more water.

Only need two, but each must have both the way and the explanation for full marks.

■ ■ ■

## Answer to part (c): average candidate response

Sustainable river management is working with a river rather than trying to control it. People are doing this now because the building of large dams and flood defences on the Mississippi does not always work, even though it costs millions of pounds.

One way is to try to put back meanders. This can prevent flooding and allow wildlife to come back. Trees planted along the river bank will stop erosion of soil and flooding. Some wetland areas can be planted with willows and reeds. Birds and water voles will then have a home. Humans will also have recreation places to visit. Problems like flooding will not be passed downstream. The cost of hard engineering is very high, so any cheaper way that does the job must have an advantage.

You can prevent new house-building near rivers, and build them higher up. The people who already live by the river would prefer the walls and dams to protect them, as they won't be able to get insured. You could make laws about what to build and where to build it, based on the risk of flooding. Farmland and football pitches go next to the river on the floodplain; housing and hospitals go further up.

Some examples of these schemes are seen on the River Skerne, where meanders have been put back, and in Florida the Kissimmee River, once drained by engineers, is being restored. There are a lot of local schemes in rivers now.

The candidate has a few basic ideas and looks at wetlands and land use zoning. The answer includes examples at the end when a case study would have been better. The candidate describes how sustainable management works but doesn't explain advantages and disadvantages. An annotated diagram or sketch could have been used. Use of terminology is limited. This is a Level 2 answer (6 marks out of 10).

■ ■ ■

## Answer to part (c): A-grade response

Sustainable management means balancing how you use something now against what you will need in the future. It is about conservation. On floodplains, important buildings like housing need to be protected, but some of these areas could be allowed to flood when necessary. Land here could be used for recreation, pastoral farming or wetlands. Cities on large rivers, like Nottingham on the Trent, have left some areas as washland, which can flood seasonally. Roads, too, are best built higher up, perhaps on river terraces. This idea is called zoning, and planning decisions are based on the likely recurrence of flooding at a particular level (e.g. 1 in 100 years).

River restoration is another idea being used on a small scale. On the River Skerne near Darlington, the old meanders, once abandoned to make the river flow faster, have been re-established. The old 'cuts' have either been kept as wetland to improve the environment and encourage flora and fauna, or can hold water in times of flood. Planting willows will not only increase interception and therefore reduce the risk of flooding, but will create a resource for the future. It has scenic, practical and ecological advantages, but will it work on a larger scale?

In LEDCs like Bangladesh, the need to use all land by the river is important and there is no money for big projects. In areas prone to flood risks, only one rice crop is grown and villagers retreat in time of flood, living and storing produce above flood level and perhaps turning to fishing. The disadvantage is that there is less food, but how many tonnes of food and lives are lost at present?

What needs to happen is for people to plan together, for the whole catchment. Dams may still be needed in some areas. The Environment Agency puts together plans for whole catchments as well as local rivers. Organisations like the RSPB also support wetland restoration. In the Netherlands, where the tradition is to use hard engineering solutions to problems, it is difficult to persuade people to try sustainable ones. However, even here, a document called *Living with Rivers* has begun to change views.

Sustainable means looking for natural solutions, which are both cost-effective and eco-friendly, where the advantages are greater than the disadvantages.

*e* This candidate has a wider view of what sustainable means and uses case studies to explain it. The answer shows a good knowledge of who and what is involved. There is good use of terminology and a logical order to the answer. Advantages and disadvantages are covered. This is a strong Level 3 response (10 marks out of 10).

# River and coastal environments

(a) Study Figure 1. It shows a river channel in the Westland district of South Island, New Zealand.

*Figure 1 A river channel in the Westland district of South Island, New Zealand*

(i) Describe the main features of the river channel shown. (4 marks)

(ii) Explain what might happen to this channel during and after a severe storm. You may annotate the diagram to show the likely changes. (6 marks)

(b) Study Figure 2. It shows a cartoon about California's 'lost' beaches.

(i) Explain how the factors shown could combine to cause very severe coastal erosion. (6 marks)

(ii) Whilst conservationists wish to 'let nature take its course', both residents and the tourist industry wish to protect the beach from erosion. Explain why. (4 marks)

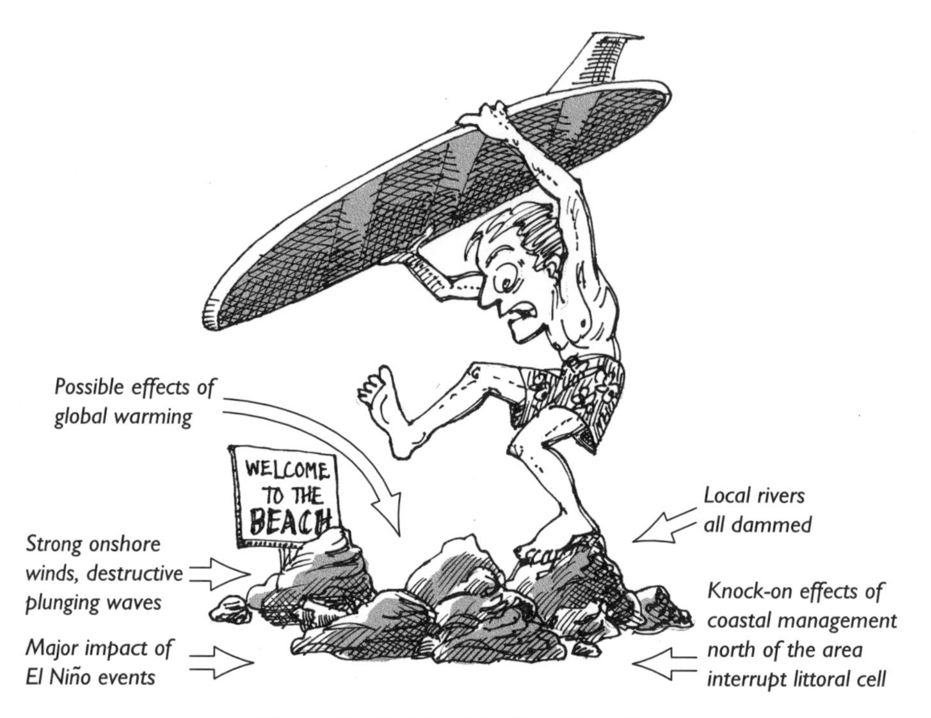


***Figure 2 California's 'lost' beaches***

**(c) A number of strategies exist to combat the problems of rapid coastal erosion. Describe *three* of these strategies and explain why they were chosen for a particular named location.** (10 marks)

■ ■ ■

## Answers to parts (a) and (b)

**(a) (i)**
- Very wide channel
- Bounded by steep sides
- Braided channel
- Some channels fast flowing; others look stagnant
- Bed covered in large stones and coarse gravel
- No vegetation, so likely to be very mobile

*e* Any four for full marks.

**(ii)**
- In flood it will fill whole valley.
- The carrying capacity will be increased.
- Larger-calibre material will be picked up and moved due to increased velocities.
- After floods, the river will take new channels.

*e* Up to 2 marks for each change explained, for a maximum of 6 marks.

**(b) (i)**
- Essentially more is being eroded than is arriving, so it is an unbalanced system — 'beach starvation'.
- Inputs are low, as dams (e.g. San Gabriel) are holding back sediment.
- Coastal management is interrupting the flow of north–south longshore drift.
- Exceptional conditions have occurred (e.g. high tides, storms, onshore winds and rising sea levels).
- El Niño events are very powerful.
- Waves are destructive as backwash is greater than swash.

**e** Up to 2 marks for the overall idea, then up to 5 × 1 for points made, for a maximum of 6.

**(ii)**
- Residents are in the front line. At the reported erosion rates, they will lose their properties in the long term.
- The tourist industry relies on there being sandy beaches. If starvation threatens the beach, erosion will also damage tourist facilities like hotels.

**e** Up to 2 marks for each view explained, for a maximum of 4.

■ ■ ■

## Answer to part (c): average candidate response

Groynes are like fences that are built down the beach at right angles, to catch the sand and stop it from being washed away. This is because if waves strike the shore at an angle, they then wash back at right angles. This keeps happening and sediments are moved along the shore by longshore drift. Groynes are usually used where the beach is in a holiday resort like Bournemouth. They are not as costly as sea walls, but do make it worse for places down the coast, as happened at Barton on Sea.

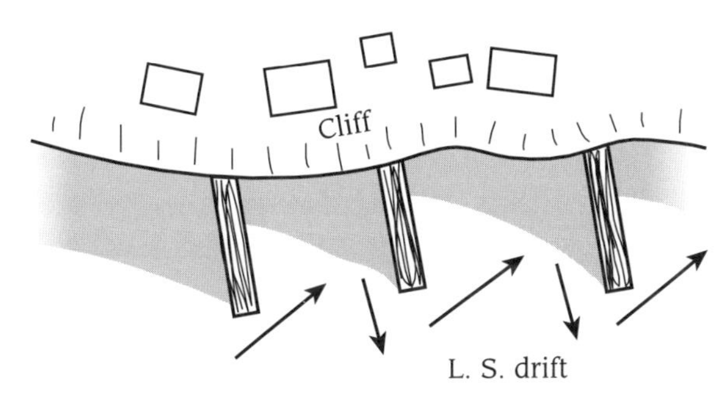


Sea walls are built along the shore to protect cliffs from erosion or land from flooding. The purpose of sea walls is that they resist the power of the waves and have to be very strong and have good foundations. Some of them have curved tops to reflect the waves. There are large sea walls at resorts like Blackpool where they protect the tower and the sea front. They are very expensive and have to be repaired often. They are usually built in places where there are buildings right on the sea front.

Beach nourishment is used to replace sand. At Miami Beach they spent millions pumping sediments from the sea floor to make a 100 m wide beach, and in 6 years it shrank to 70 m. This can only be used where the benefits of a beach are large, like the hotels and tourist facilities of Florida. You also need the money and technology to do it.

**e** These three strategies are correct factually and each is described, but the explanations are not convincing. The examples chosen are widespread, rather than being from a study of a single stretch of coastline. This is quite a sound Level 2 answer (6 marks out of 10).

■ ■ ■

## Answer to part (c): A-grade response

On the Holderness coast of Yorkshire there are a number of different strategies being used. The strategies were chosen because this is a very exposed coast that is mostly made of boulder clay which erodes easily. There is also very little beach to protect it.

At Mappleton the beach had been eroded by waves and longshore drift, making the cliffs retreat at over 1 m per year. The clay from the slumping cliffs is being removed in suspension. To halt this process two groynes (together with some rip-rap and beach nourishment) were built in 1991 at a cost of £2.1 million. These have trapped the sand and reduced cliff erosion, so the small cliff-top village and its coast road look secure. A farm at Cowden Beck, down the coast, has, however, been eroded from the cliffs there. Groynes were used because they seem to be cost-effective for relatively small schemes.

Further south along this coast is Withernsea, a small holiday resort. Here, in winter, the waves have until recently washed over the old sea wall to the hotels behind. Now a new sea wall has been built with a wave return top. This is designed to reflect as well as resist the effects of storm waves from across the North Sea. In addition, it has been fronted by a line of rip-rap boulders which not only help to absorb the power of the waves but prevent scouring of sand. The decision to build a wall may reassure local businesses, residents and even tourists, even though it was at a cost of £6 million.

At Spurn Head at the southern end of this coast the spit is very fragile. At the neck, where the spit is narrow and near to sea level, storm waves can wash over the sand dunes. In 1995 the local council decided to abandon repairs here as the £20 million needed to make the area safe was too high a price. The small community living on the Head are hanging on, repairing the roadway themselves. This policy is really just 'doing nothing'. Naturalists and Heritage coast managers are quite happy with this decision.

**e** These strategies are well supported with evidence about their nature and why they were built at each location. The structure of the answer is excellent and the candidate displays clear understanding. This is a sound Level 3 answer (10 marks out of 10).

# Coastal environments

**(a) Study Figure 1. It shows a sketch of a section of coastline.**

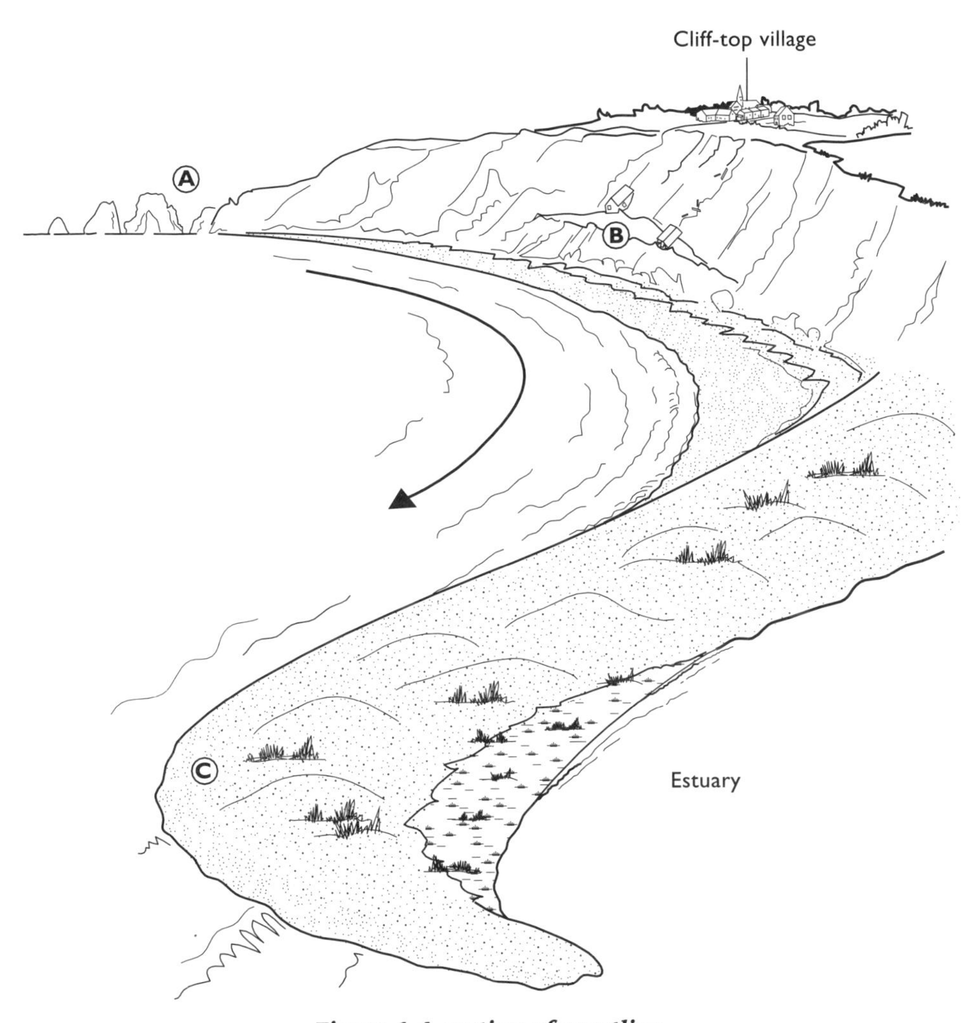


*Figure 1 A section of coastline*

**(i) Identify the coastal features at A, B and C.** (3 marks)

**(ii) Suggest how the features at A might have formed.** (5 marks)

**(iii) Villagers living above the cliff at B are told that their homes are threatened by cliff-foot and cliff-face processes. Explain the differences between these two types of process.** (2 marks)

**(iv) Suggest what factors may have helped to shape the depositional feature at C.** (3 marks)

**(v) Explain why the features A, B and C form part of an *open system*.** (2 marks)

**(b) Study Figure 2. It shows a cross-section through the feature at C.**

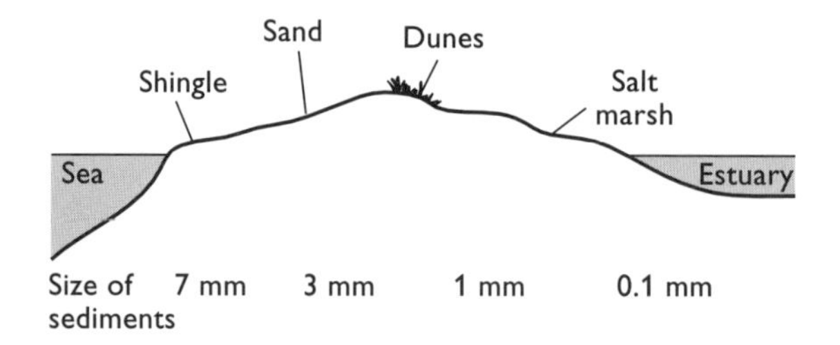


***Figure 2 Cross-section through the feature at C***

**Describe and suggest reasons for the changes in sediments and vegetation seen across this environment.** (5 marks)

**(c) For a named coastal environment, examine the impacts of *one* of the following:**

- **urban and industrial development**
- **recreational and tourism pressure**
- **land reclamation, linked to agriculture** (10 marks)

## Answers to parts (a) and (b)

**(a) (i)** A — Stack, arch or headland
B — Landslide or slump
C — Spit (recurved)

1 mark for each.

**(ii)**
- Combination of wave action and rock lithology
- Sequence of events from headland, cave, arch and stack
- Wave erosion by abrasion (throwing material at cliffs)
- Wave quarrying (hydraulic action trapping air in cracks)
- Lithological weaknesses, such as joints or faults in the rocks

1 mark for each, based on linking features to processes.

**(iii)** Cliff foot — marine processes that attack the beach or base of the cliffs (e.g. wave quarrying)
Cliff face — sub-aerial processes that attack the cliff face (e.g. slumping)

1 mark for each.

**(iv)**
- Change in line of coast
- Change in dominant wind direction
- Longshore or tidal currents
- Effects of estuary outfall

1 mark for each, up to a maximum of 3.

(v) • Not balanced
• Additional input from estuary
• Losses down the coast by longshore drift

e 1 mark for each, up to a maximum of 2.

(b) • Sediments change in size — shingle/sand/mud — due to change in energy and origin (strong waves/wind/quiet water etc.)
• Vegetation changes in type — none/dune/saltmarsh — due to change in environmental factors (shelter, salt, water depth, etc.)

e 2 × 1 marks for the descriptions and 2 × 2 marks for the reasons, up to a maximum of 5 marks.

e **Parts (a)(ii) and (b) could be marked using Levels.**

■ ■ ■

## Answer to part (c): average candidate response

Urban and industrial development can cause a lot of damage to a coastal environment. Teesside on the coast of northeast England is a major centre for chemicals. Big companies like BSC (steel) and ICI brought many jobs and these in turn led to others. They also brought pollution, especially in the Tees Estuary. However, the bad days of the 1970s have gone with the river 70% cleaner now, though heavy metals still pass down river. Nitrogen causes eutrophication and oil spills harm birds. Fish stocks are declining and this is probably linked to pollution.

Cities like Middlesbrough and Hartlepool surround the estuary and demand more space. Coastal centres like Redcar are now becoming commuter towns as well as holiday resorts. This increases traffic and is using up more of the area's open space. Wetland to the north is being reclaimed by draining and filling it in. Domestic sewage is also increasing. The Teesside Development Corporation has tried to promote development in the area with large plans like the Tees Barrage, which it was hoped would improve the economic development of the area. Environmental groups like English Nature say it will affect the river and the water table.

e This answer needs a better framework. Some good points are made about estuaries, and Teesside is given as an example. However, the answer is very much about industrial development and there is certainly not enough about urban impacts on coasts. This is a low Level 2 answer (5 marks).

■ ■ ■

## Answer to part (c): A-grade response

Tourism is Dorset's most important industry. The season is short and is busiest from June to August. Along the coast, there are natural features like Lulworth Cove. Many recreation activities are water-based and these have grown very rapidly — for instance around Poole Harbour. Inland, the marketing of Wessex and Thomas Hardy have widened the area's appeal. In addition to all the facilities gained, income in the region is now worth over £¾ billion, with Bournemouth as the premier resort.

However, not all impacts are quite so beneficial. Socially there are worries about the quality of bathing beaches and the pressure on water supplies in some summers. Out of season, many facilities close and local residents say visitors bring higher prices, noise and anti-social behaviour. Economically, many jobs are poorly paid and part-time, as well as seasonal. Second homes push up rents for younger locals, and foreign travel means that less money is spent there.

Environmental impacts are where the real problems lie. Traffic congestion in towns like Weymouth, on main roads like the A35 and in honeypot locations, is increasing. Parking is an almost insoluble problem. Unsightly caravan parks, inappropriate developments and footpath erosion of cliff-top walks all need to be better controlled.

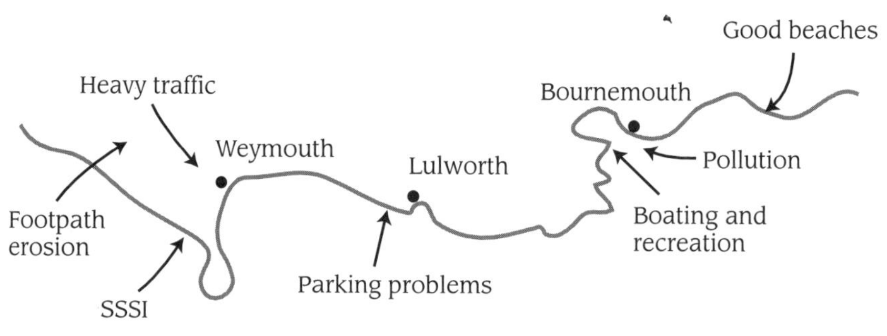


This seems a very negative view, but a more sustainable strategy is possible. The Ferrybridge management plan near Chesil Beach, a honeypot location, has put in place a variety of solutions. These include identifying SSSI and RAMSAR sites, building proper boardwalks across the Hamm and providing windsurfers with a convenient car park. The answer for the area as a whole is to try to assess what the environmental impacts are going to be and have a more integrated approach to managing tourist pressures

*e* There is a nice structure to this essay and, while it is short on facts, it is well written and shows a clear understanding of the issues. This just fails to reach Level 3 and is awarded a top Level 2 mark of 8 out of 10. More facts would have taken it to Level 3. An A-grade candidate is likely to achieve a mixture of Level 3 and good Level 2 answers.